802.1aq Shortest Path Bridging
Design and Evolution

802.1aq Shortest Path Bridging
Design and Evolution

The Architect's Perspective

David Allan

Nigel Bragg

Published by
Standards Information Network
IEEE Press

A John Wiley & Sons, Inc., Publication

Contents

Contents

Figures

Acknowledgments

This document is an amalgam of practical wisdom of Dave Allan, Peter Ashwood-Smith, Nigel Bragg, Janos Farkas, Don Fedyk, Jérôme Chiabaut, Dinesh Mohan, Mick Seaman, and Paul Unbehagen.

Thanks also to Anne Bragg for her persistence in reducing our wayward capitalization, punctuation, and other stylistic idiosyncracies to more conventional forms.

Technical Reviewers

Simon Parry, *Ciena Limited*
Joel M. Halpern, *Distinguished Engineer, Ericsson*

Introduction

Ethernet is a difficult and demanding taskmaster.

We start from the position that for any networking technology of sufficient power, an elegant and self-consistent solution to a given connectivity problem exists. Switched Ethernet is the product of 30 years on Occam's razor, and although the attributes and scale of the application domains covered by this book were until recently inconceivable, we have concluded that it remains a technology of sufficient power and self-consistency.

The success and longevity of Ethernet can be put down to the fact that it has been able to evolve to accommodate new requirements, both in its original LAN application space and in the increasing proportion of Provider networking space. Shortest path bridging (SPB) is one of the most recent of these evolutionary steps, and we would like to establish at this early point both what is the fundamental problem it solves and why the solution is useful.

The short and sufficient answer is, "elimination of the Spanning Tree Protocol and its shortcomings, and its replacement by a superior routed technology, and without changing the service model." This answer is "sufficient" for now because it is generally accepted in the industry that Spanning Tree Protocol presents problems and limits the applications accessible to Ethernet, and we therefore defer further discussion on the origins and root cause of this problem to the beginning of the next section.

Replacement of Spanning Tree Protocol by something substantially superior is a general "good" that applies to Ethernet networking in both Enterprise and Provider space. The other key requirement of Ethernet networking, which is increasingly shared by Enterprise applications as well as Providers, is virtualization, which is the ability to support multiple independent LAN segments on the same physical infrastructure. SPB did not originate the technology to do this, but directly supports earlier IEEE Standards (Provider Bridging and Provider Backbone Bridging) that defined the hierarchical data path constructs to support virtualization.

There are two variants of SPB, one using the 802.1ad Q-in-Q datapath—shortest path bridging VID (SPBV)—and one using the hierarchical 802.1ah MAC-in-MAC datapath—shortest path bridging MAC (SPBM). SPBV and SPBM share a control plane, algorithms, and common routing mechanisms; where the term "SPB" is used, this describes aspects common to both variants.

The authors embarked on their journey motivated by the issues of highly scalable networks intended for deployment by Service Providers, a path that lead to a precursor of SPBM, known as Provider Link State Bridging, or PLSB. A significant number of the topics discussed are more relevant to a technology supporting virtualization, and the reader should therefore expect a significant focus on SPBM, both because of its highly scalable support of virtualization and, as a corollary, because of from where the authors came.

The first, and substantially shorter, part of this book summarizes succinctly and informally what SPB "is" today, with the aim of offering a reader new to the technology a consistent mental model of what it does and how it does it. The second part provides the rationale for *why* SPB is as it is, and has to be so. This therefore not only includes a post hoc rationalization of SPB with the 20:20 vision of hindsight, but also describes some of the blind alleys explored in getting there, because these alleys give additional insight into why SPB has to be what it is.

We start with a short history of SPB and its antecedents, with only the briefest allusions to the motivations for SPB at this stage (Chapter 1, "IEEE 802.1aq in a Nutshell; Antecedents and Technology"). We then offer a short description of SPB as it is now (three sections on "SPB technology: The Control Plane," starting on p. 15). These are succinct, but capture the key principles and attributes of SPB. We nonetheless anticipate that readers will finish this with more questions than answers, such as:

- Why *is* congruence so important?
- How big a network can you really make?
- Why is this really so different from other network technologies?

The rest of the book sets out to answer these questions.

We start by considering the key requirements that the different networking scenarios present (the section on "The Problem Space,"

p. 37). As a way of introducing the constraints and degrees of freedom offered by the Ethernet baseline, we follow this with a summary of our progress toward SPB as documented here ("History," p. 52). Unlike subsequent chapters which organize material by topic, this is a chronological record that talks through some of the twists on the journey.

We then revisit some fundamental principles of Ethernet, showing why we decided to stick with some key bridging constructs even though the adoption of a control plane meant that they were no longer strictly mandatory. We also reinterpret the use of some other key Ethernet concepts, such as the Virtual LAN (VLAN), always within their strict specifications, but in ways possibly contrary to received wisdom on their usage.

In the section on "Rounding Out Design Details" (p. 69), we focus on overall networking challenges beyond basic functionality that have to be addressed to make a technology deployable. There is a discussion of data plane instrumentation, the OAM (deliberately short, to make an important point), followed by more extensive discussion on dual-homing for resiliency, always a thorny issue for Ethernet.

The section in Chapter 3 on "The Control Plane Is as Simple as It Can Be, but No Simpler" (p. 74) shows that essentially all SPB functionality can be delivered by the routing system. We first discuss SPB's most radical departure from previous received wisdom, the complete elimination of signaling from both unicast and multicast state installation. We then provide a factual introduction to the extensions to IS-IS required by SPB, showing how modest these are. Finally, we explain some of the algorithmic innovations required by SPB over previous link state routing practice.

So far, the exposition has assumed point-to-point connectivity between bridges in the SPB domain, and ignored the traditional shared segment. Chapter 4, "Practical Deployment Considerations" (p. 130) considers this and other topics, because an SPB overlay of an *emulated* LAN segment is a real deployment scenario. Although a solution is described, this is not quite a "done deal," because it needs modest extensions to the Ethernet forwarding path.

Next, in Chapter 5, we explore applications of SPB (p. 150), providing walk-through examples of operation in various deployment scenarios covering the delivery of Metro Ethernet Forum defined services by carriers, and the use of SPB in enterprise applications. Because Ethernet has been very widely deployed in these applications in the past, this treatment focuses on the ability of SPM to address the limitations and deficiencies of earlier Ethernet technologies.

Finally, in Chapter 6, we explore whole new capabilities that SPBM could be extended to provide.

In SPBM, the service primitive is the emulated LAN segment. The LAN segment at Layer 2 *is* the IP Subnet at Layer 3, and IS-IS has been routing IP for many years. So, if IS-IS retains its IP personality as well as running SPBM, we have a single control plane with a complete view of both Layer 2 and Layer 3 topologies, and we can "route at the edge, switch through the core," and virtualize the notion of location implied by the subnet prefix. The SPBM service architecture can now be used to construct a virtual *network* of IP Subnets, and the result is *IP-VPN* capability as well as the native virtual LAN segments. This is elaborated in the section on "Layer 3 Integration with SPBM."

We also investigate how the "Multiarea" capability of IS-IS can be applied to SPB. Finally, we explore how the shortest path tree may be extended with other connectivity styles and incorporated into the framework supported by the control plane (the section on "Extended Connectivity Models: Spanning Trees"). We first show how traditional spanning trees may be constructed. We then turn our attention to the coercion of traffic off shortest paths, for traffic engineering purposes, without causing undesirable side effects within the routed system.

We started by asserting that the success of Ethernet and its evolution is the consequence of 30 years on Occam's razor and that we have discovered that it is a "technology of sufficient power and self-onsistency." We hope the reader in the process of the journey of discovery outlined above ultimately agrees with us.

As an early hint as to *why* we believe this, Ethernet descended from a broadcast medium. This is very important, as the types of connectivity offered by Ethernet are derived from filtering of the basic broadcast behavior, with the point-to-point connection simply being the most extreme form of filtering. The implication here is that all types of communication—one-to-all, one-to-some, and one-to-one—can be derived from the basic transmission behavior combined with filtering. This is distinctly different from the history of most other network technologies, which have started from one-to-one connections as the service primitive, and subsequently overlaid broadcast behavior onto this unicast model.

We also attach much importance to the fact that Ethernet uses *global addressing* in the data plane, a characteristic that it shares with only one other major production technology, IP. Everything seems to follow from this choice, rather than the adoption of a "link-local" identifier as a forwarding scheme:

- with the proper control plane it scales. Scalability is in practice dominated by state volume, and not theoretical considerations of addressing space; IP uses address aggregation to control the issues of global addresses; Ethernet uses *hierarchy*, the alternative route to scale.
- global data plane identifiers make a frame self-describing and remove the need for rafts of complexity; signaling can be eliminated because global identifier information can be communicated by more efficient means; whole classes of subtle errors caused by lack of synchronization between control and forwarding planes are eliminated; and the OAM to detect the remaining fault classes is much simpler.

Ethernet continues to evolve, and this book is simply a snapshot of a point on the journey. We have endeavored to provide insight into what we believe to be a significant evolutionary step in Ethernet technology. Ethernet's longevity and its ability to evolve to address new requirements are an independent testimony to its fundamental "fitness for purpose." However, as it has evolved, the limitations of spanning tree have become increasingly apparent, and finally become a real barrier to further extensions to its scope. With SPB, Ethernet has acquired the state of the art in distributed routing technology, which is now available to future evolutionary developments.

Abbreviations

ABB	Area Boundary Bridge
AESA	ATM End System Address
ARP	Address Resolution Protocol
ATM	Asynchronous Transfer Mode
BCB	Backbone Core Bridge
BEB	Backbone Edge Bridge
BGP	Border Gateway Protocol
BNG	Broadband Network Gateway
BRAS	Broadband Remote Access Server
CLIP	Classical IP over ATM
CSNP	Complete Sequence Number Packet
CTO	Chief Technology Office/Officer
DA	Destination MAC address
DSLAM	Digital Subscriber Loop Access Multiplexer
ECMP	Equal Cost Multi Path
ECT	Equal Cost Tree
EMS	Element Management System
ESP	Ethernet Switched Path
IETF	Internet Engineering Task Force
IGMP	Internet Group Management Protocol
IP	Internet Protocol
I-SID	I Component Service ID
IS-IS	Intermediate System to Intermediate System
ITU-T	International Telecommunications Union—Telecommunications Standardization Sector
IVL	Independent VLAN Learning
LAN	Local Area Network

LDP	Label Distribution Protocol
LSP	Link State Packet
LTM	Link trace message
MAC	Media Access Control
MEF	Metro Ethernet Forum
MEL	Maintenance Entity Level
MEP	Maintenance End Point
MIP	Maintenance Intermediate Point
mp2p	Multipoint to point
MPLS	Multiprotocol Label Switching
NHRP	Next Hop Resolution Protocol
NNI	Network to Network Interface
NSAP	Network Service Access Point
OAM	Operations, Administration and Maintenance
OLT	Optical Line Termination
OUI	Organizationally Unique Identifier
p2mp	Point to Multipoint
PBBN	Provider Backbone Bridged Network
PBT	Provider Backbone Transport
PLSB	Provider Link State Bridging
PSNP	Partial Sequence Number Packet
RPFC	Reverse Path Forwarding Check
RT	Route Target
SA	Source MAC address
SDH	Synchronous Data Hierarchy
SONET	Synchronous Optical Network
SPB	Shortest Path Bridging
SPBM	Shortest Path Bridging MAC Mode
SPBV	Shortest Path Bridging VID mode
STP	Spanning Tree Protocol
SVC	Switched Virtual Circuit
TRILL	Transparent Connection of Lots of Links
TTL	Time to live
UNI	User to Network Interface

VID	VLAN ID
VLAN	Virtual LAN
VPLS	Virtual Private LAN Service
VPN	Virtual Private Network
VSI	Virtual Switch Instance

CHAPTER 1

IEEE 802.1aq in a Nutshell: Antecedents and Technology

The Enterprise Local Area Network (LAN) is the traditional Ethernet domain. However, Ethernet has throughout its history widened the range of applications and markets that it could address. Now it is increasingly being equipped to address the provider space, which has significantly different requirements, notably the capability to virtualize large numbers of services to run on common infrastructure. These requirements were the initial motivation for IEEE 802.1aq—Shortest Path Bridging (henceforth SPB).

SPB: ANTECEDENTS AND PRINCIPLES OF NETWORK OPERATION

Summary of Ethernet Connectivity Models

Ethernet was invented to deliver LANs, offering "plug and play" networking, and required no configuration in its original form. Addresses are burned into endpoints at manufacture and are not under the control

802.1aq Shortest Path Bridging Design and Evolution: The Architect's Perspective,
First Edition. David Allan and Nigel Bragg.
© 2012 the Institute of Electrical and Electronics Engineers. Published 2012 by John Wiley & Sons, Inc.

of the network. The LAN was a passive medium (coax cable), and a collision detection mechanism was used to arbitrate access to this single shared medium by multiple endpoints. Broadcast was the only native connectivity type, with endpoints being responsible for filtering frames which were not addressed to them. Over time, the requirement emerged to scale Ethernets beyond the 6000-foot limit imposed by the collision detection mechanism. This resulted in the development of bridging, and with it the need to discover the location of endpoints across the bridged network.

This was arguably the only major architectural discontinuity in Ethernet's history, the transition from a LAN segment implemented as a passive shared medium to an actively switched network. This transition was achieved while preserving unaltered the service offered to clients, but it required a completely new network technology, the learning bridge. To allow this perfect emulation of a passive shared medium, the routing system adopted by bridged Ethernet then, and still specified, is flood-and-learn; frames with destination media access control (MAC) addresses unknown to intermediate switches are flooded, and the correct port to use for forwarding subsequent unicast traffic back to the source of the flooded frame is found from the source address by reverse path learning. To permit such broadcast mechanisms without frame looping and network meltdown, the active topology must be highly constrained and offer symmetric connectivity between any two points, the common case being a simple spanning tree.

The moment multiple paths between switching points are installed in this bridged model, whether deliberately for resiliency or acciden-

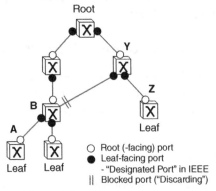

Figure 1.1   A simple spanning tree.

tally by misconnection, a loop is created, which results in network meltdown when broadcast is used. The Spanning Tree Protocol (STP) was developed to disable all redundant paths in a bridged network and create an active topology which is a simple spanning tree (with only one path between any pair of endpoints), and which therefore appears as an exact replica of the coaxial shared medium.

Figure 1.1 illustrates the salient attributes of spanning trees. The tree itself is a directed graph from the root node, which is typically administratively determined, but it is important to realize that the forwarding path thereby established is bidirectional. This means that "go" and "return" paths between any two endpoints must be congruent, which is fundamental to the traditional Ethernet "flood and learn" data path routing process.

The first time a frame, destined for an endpoint attached to Z (say), is sent by a source attached to A, it is flooded by A as "unknown." Copies of the frame traverses the entire tree, and its port of arrival on intermediate bridges allows them to learn how to reach the original source, with no other knowledge of the network at all; on a simply connected tree, a reply can only be delivered by returning it through the port through which the original message arrived. The same mechanism allows the reply from the endpoint attached to Z to teach the intermediate bridges which port to use to reach Z on subsequent occasions.

These mechanisms work functionally and robustly, but have undesirable consequences:

- all redundant links, representing real dollar investment, are turned off; in Figure 1.1, the link between B and Y must be turned off to prevent the formation of the obvious loop.

- as a consequence, traffic routing is often suboptimal, in particular for traffic between leaves having disjoint paths to the root; in Figure 1.1, traffic between A and Z is not able to take the "shortest path."

- the spanning tree offers simply connected connectivity to the set of endpoints served and hence is a single point of failure; the need to guarantee lack of loops at all times requires **all** connectivity on a spanning tree to be disabled after **any** topology change until the new tree has converged. Originally this "shutdown" period typically lasted tens of seconds; this has been improved, but recovery dynamics are still regarded as unacceptable.

SPB introduces link state routing to Ethernet to replace the distance vector algorithm underlying STP, and uses sets of shortest path trees in lieu of a single or small number of spanning trees. This addresses both issues cited above:

- with full topology knowledge, link state allows the control plane to construct loop-free shortest path trees, with no need to disable any data plane connectivity;
- the use of per-source shortest path trees means that connectivity unaffected by a topology change is uninterrupted;
- link state routing inherently has much better convergence properties than distance vector, and SPB has further improved these with speed-up techniques which exploit Ethernet's innate multicast properties.

This replacement of STP by something substantially superior is a general "good" which applies to Ethernet networking in both enterprise and provider space.

Introduction to Virtualization Support in Ethernet

The other key requirement of Ethernet networking, which is increasingly shared by enterprise applications as well as providers, is virtualization, which is the ability to support multiple independent LAN segments on the same physical infrastructure. SPB did not originate the technology to do this, but directly supports earlier IEEE Standards (Provider Bridging and Provider Backbone Bridging) which defined the hierarchical data path constructs to support virtualization.

To provide a summary of virtualization support by Ethernet, Figure 1.2 shows evolution of the increasingly rich header formats which have been defined. In this, a hierarchical layering is implied by prefixes associated with the well-known terms MAC (often used as shorthand for MAC address) and VID (virtual LAN [VLAN] identifier). The prefix "C" refers to customer address information, the prefix "S" refers to provider imposed tags (VIDs) in a Q-in-Q network, and the prefix "B" refers to backbone address information in a MAC-in-MAC network.

A major contribution of SPB is to offer a replacement for spanning tree that is capable of fully utilizing much more richly connected

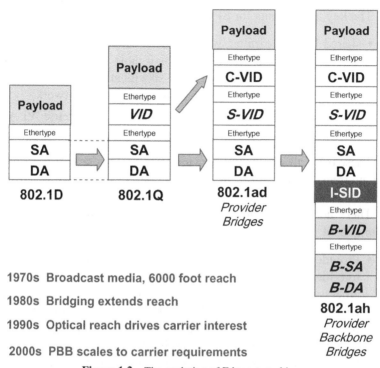

Figure 1.2   The evolution of Ethernet stacking.

topologies. SPB is an umbrella term covering two modes of operation:

- SPBV is VLAN based and builds upon the IEEE 802.1ad (Q-in-Q) tagging structure of Provider Bridges to construct shortest path trees each defined by a different VID,
- SPBM in which the shortest path trees are MAC based (the B-MAC space of Provider Backbone Bridges). VLANs are used to delineate multipath variations.

At a 50,000-foot level, both SPB modes use very similar operations and have very similar overall properties, the primary differences emerging as a consequence of the scaling limitations of the SPBV data plane, which are not shared by SPBM.

The key antecedents for SPBV are:

- shared VLAN learning, whereby MAC addresses learned in one VLAN populate a shared forwarding table for a set of VLANs; this development came with the specification of the ability to support multiple instances of spanning tree in a network;
- the concept of unidirectional VLANs or asymmetric VLANs [SPB].

These two collectively permit a properly constructed mesh of shortest path trees constructed from unidirectional VLANs to employ traditional flooding and learning outside a spanning tree context.

There are two key antecedents of SPBM which foreshadow the techniques it uses:

- Provider Backbone Bridging (IEEE 802.1ah PBB) introduced true hierarchy to Ethernet for the first time, with customer Ethernet traffic using what are referred to as C-MAC addresses being encapsulated in Backbone MAC (B-MAC) addresses across the backbone. This has a number of benefits; the key ones to be aware of now are that the hierarchy directly supports virtualization, and that **all MAC addresses in the backbone are known to and in control of the network operator;** C-MACs are encapsulated and therefore hidden, and B-MACs are all associated with switches in the PBB network itself. This offers a significant degree of summarization of state across the backbone.
- PBB has a comprehensive architectural model, which defines nodal roles in relation to the backbone network (known as a Provider Bridged Backbone Network or PBBN). These are the Backbone Edge Bridge (BEB), which is a node that has both UNIs and NNIs, and the Backbone Core Bridge (BCB) which is purely a transit device at the backbone layer.
- PBB-Traffic Engineering (802.1Qay PBB-TE) exploited this complete knowledge of backbone addressing and topology to permit the disabling of Ethernet's native routing system—flooding and learning. Instead, forwarding tables were **explicitly populated by management or a control plane.**

SPB also explicitly configures forwarding tables, but uses a different control regime, and we now introduce this.

Introduction to Path Computation in SPB

SPB takes a radically different approach to the construction of connectivity compared with spanning tree, but with the result consistent with Ethernet principles. SPBV constructs shortest path forwarding trees between all Provider Bridges in an SPBV domain using shortest path VIDs (SPVIDs) to identify each tree. It is required that the go and return paths between any two bridges, identified by their respective SPVIDs, share a common route in order that source learning works across the core. This permits the "flood and learn" paradigm of bridging to be retained while keeping endpoint state out of the control plane. SPBM constructs shortest path forwarding trees between all BEBs in the network using the combination of B-VIDs and MAC addresses. The MAC learning process in the B-MAC layer is adapted to become a frame-by-frame policing of loop freeness. Symmetrical metrics are used to ensure unicast/multicast congruency and bidirectional fate sharing, both highly desirable properties for Ethernet services. In both cases, the information to derive the forwarding databases is distributed by a link state routing system.

Shortest path forwarding within the Ethernet architecture is achievable because it is possible to fully connect a network with shortest path trees such that there is bidirectional symmetry of the forwarding path between any two points in the network. MAC (SPBM) and VID (SPBV) entries in the forwarding tables are populated by the control plane. This requires placing the responsibility for maintaining a loop-free active topology on handshaking within a link state control plane, and moving away from Ethernet's traditional reliance on a strictly maintained and simply connected spanning tree in the data plane.

Furthermore, it is both possible and practical to condense all SPB control and configuration into a single control protocol: Intermediate System to Intermediate System (IS-IS), which is fundamentally a robust means of synchronizing a common repository of information across multiple platforms. This consolidation is possible because the VID (SPBV), also the Provider B-MAC, B-VID, and Service Identifier information in the form of the I-SID (SPBM) is all global to the network, and so link local forwarding state (e.g., Frame Relay data link connection identifiers, or MPLS labels) is not required for SPB. In other words, the SPB control plane has no need to describe the modification of identifiers within link state control packets crossing the network, because this identifier information is invariant across the network, which is the exact corollary to the fact that in the data plane, Ethernet frames transit the network unmodified; neither requires personalization. Consequently, small extensions to the IS-IS protocol permit control plane flooding of the required VID, B-MAC, and I-SID information within the network.

Connectivity is constructed using a distributed routing system where each node independently computes the local filtering database (FDB), used for the actual forwarding of frames, from the information in the routing system database. The necessary personalization exists only in the form of each node's local view of its position in the network which is extracted from a common information repository during the FDB generation process.

A converged network can have numerous fully connected multipath solutions implemented in the data plane; for SPBV, this requires consuming a VID per node per solution in the network, and for SPBM, one solution can be instantiated per B-VID. So while in the case of SPBV the size of the network directly affects the number of potential multipath solutions that can be deployed, for SPBM it is independent of network size, and the design limit is based on the limitations of the B-VID code space, with in theory support for 4094 multipath solutions before the identifier space is exhausted.

The configuration of a single fully connected solution in a converged network will typically have a multipoint-to-point (mp2p) unicast tree **to** each node in the network (shown as solid black lines to node "A" in Fig. 1.3), and a congruent point-to-multipoint (p2mp) broadcast tree **from** each node to its peers (the pecked lines from node "A" in Fig. 1.3), the latter performing what [Metcalfe] referred to as "reverse path forwarding." These are constructed such that the point-to-point (p2p) path between any two points in the network in a given multipath solution is symmetric and congruent in both directions, and this is true for both unicast and multicast. The tree rooted on node "B" (shown in dotted lines in Fig. 1.3) shows that its shortest path connectivity is different from other trees, but the connectivity between "A" and "B" always uses the reverse of the path from "B" to "A." For SPBM, each p2mp broadcast tree is the prototype for construction of "per I-SID" (per service) multicast trees pruned to connect only BEBs participating in a specific I-SID. The set of multicast trees built to support a specific I-SID offer a perfect virtualized emulation of a traditional Ethernet LAN segment.

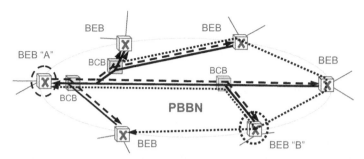

Figure 1.3   IEEE 802.1aq technology—data plane connectivity for BEB "A."

The SPBV Data Plane

Provider Bridging was the first product of the endeavor to adapt Ethernet technology to carrier needs. Provider Bridging specified the imposition of an additional "outer" VLAN tag to the Ethernet frame, known as the S-tag, which permitted a provider to both isolate a customer in a provider network, and also to control subsetting of the network connectivity (the traditional function of a VLAN), in order to properly implement a service instance as a bridged closed user group (see the stack shown in Fig. 1.2).

Embodied in 802.1ad is the ability to translate VLAN tags at provider boundaries to ensure providers can independently administer their own tag spaces, and so avoid a reassignment in one domain propagating into other domains. SPBV also uses this capability. It had never been part of Ethernet before 802.1ad, because a VLAN describes a network-wide topology, and moving between VLANs requires routing, not bridging. It is important to understand this very specific meaning of the VLAN and its tag, which is quite different from the link-local "identifier" used by MPLS, and Asynchronous Transfer Mode before it, where label swapping on every hop is a fundamental part of forwarding. The Ethernet VLAN tag translation function is normally a symmetric function where tag "A" is translated to tag "B" in one direction, and tag "B" is translated to tag "A" in the other.

SPBV implements a VLAN using a set of unidirectional shortest path VIDs (known as SPVIDs), each being used by a different Provider Bridge to mark frames which it transmits into the SPBV network. It is necessary to have an identifier to refer to the complete set of SPVIDs implementing a VLAN, and this is known as the "Base VID" in IEEE documentation. The Base VID is the VLAN that a set of SPVIDs implements.

The tag translation functionality of 802.1ad needs to be modified for SPBV. A VLAN tag received at the ingress to an SPBV network will map to a single SPVID associated with the ingress Provider Bridge, but the reverse function needs to translate the complete set of N SPVIDs associated with the VLAN, one from each Provider Bridge, back to the single S- or C-tag value for egress from the SPBV network. This has necessitated the extension of the symmetrical tag translation concept above for use by SPBV. This defines an N to 1 tag mapping on egress from an SPBV domain, but continues to exclude tag swapping as a switching function (i.e., there is still no 1 to N mapping needed because the mapping on ingress to the domain is 1 to 1).

An SPVID is unidirectional, and the set of SPVIDs that implements a VLAN operates in shared VLAN learning (SVL) mode. Within a bridge, a VLAN, and its tag, are associated with the use of a single FDB to control frame forwarding on that VLAN. With shared VLAN learning, multiple VLANs are assigned to the same FDB. As a consequence, a MAC address learned when received as a source address on one VLAN is used when received as a destination address on any VLAN to determine the forwarding action.

In the specific case of SPBV, a source MAC in a given frame, tagged with the SPVID of the ingress SPBV provider bridge, is learned by all SPBV bridges transited by the frame as applicable to the complete set of SPVIDs associated with the VLAN the MAC arrived on, so that a frame destined for that original source may be forwarded irrespective of the bridge (hence SPVID) which sent it.

Existing MAC registration protocols for multicast groups may interoperate with an SPBV environment, and registrations received at the edge of an SPBV region are advertised throughout the region using IS-IS.

The SPBM Data Plane

Provider Backbone Bridging (IEEE 802.1ah PBB) was the culmination of the evolution of the Ethernet forwarding path, allowing for a full encapsulation of the customer functions of topology and service identifying frames. SPBM inherits this forwarding path unaltered. Both PBB and SPBM use an 802.1Q standard header and an S-VLAN

Ethertype, but unlike Provider Bridging, separate the service identifier from the backbone VLAN (B-VLAN) and instantiate it completely independently as the I-SID (see Fig. 1.2). This is important since the number of VLAN topologies is typically a scaling constraint for Ethernet (only 4094 VIDs are available), and so when the VID is overloaded and used as a service identifier as well, this severely impacts the number of services a Provider Bridged Network can support.

The separation of VID and service ID permits the services to scale independently of topology; the B-VID is then delegated exclusively to the role of engineering the network. SPBM also uses the term Base VID (as above) to refer to the VID identifying the VLAN, but unlike the case of SPBV where an SPVID must be used to identify the source bridge, SPBM can use the edge bridge B-MAC address for this purpose. This is because the domain of the IS-IS control plane is fully congruent with the set of endpoints in the backbone. Consequently, SPBM can fully mesh the network with a single VID, and so there is a 1:1 correspondence between the Base VID and the SPBM B-VID.

The I-SID is a service identifier which is unique and consistent within a provider network. The binding of a particular I-SID to a set of BEB customer network ports uniquely identifies a community of interest, which is implemented as a virtual switched broadcast domain between those ports, over which customer transparent bridging operates. I-SIDs are normally associated with a single B-VID.

Customer Ethernet traffic is adapted onto an SPBM network in the same manner as used in 802.1ah PBB. A customer's Ethernet frame arrives at a BEB at the edge of the SPBM network, and is mapped to the customer I-component and I-SID associated with the customer tag or port. Associated with the I-component is a table, exactly analogous to the FDB of a physical bridge, which records the set customer MAC addresses received together with the B-MAC address of the remote BEB which encapsulated and sent them. This is exactly the normal reverse path learning mechanism associated with bridging, except that C-MAC addresses are here associated with the B-MAC address of the BEB via which the C-MAC can be reached, rather than a physical port. Thus, the backbone MAC simply becomes a named interface in a large distributed bridge.

When the customer frame's destination C-MAC cannot be resolved to a B-MAC, or it is a broadcast or multicast frame, then the I-component will resolve the address to a Group (multicast) B-MAC associated with the local I-component, which identifies the specific shortest path multicast tree over the backbone for the combination of that source and I-SID.[1] Forwarding of the frame addressed in this way floods the frame to the other BEBs that have registered interest in receiving that I-SID. The peer BEBs learn the customer source C-MAC to ingress BEB B-MAC binding, analogous to MAC learning today but using B-MAC named "ports" rather than physical ones. When a response is elicited from the customer destination, the initial ingress BEB learns the binding of C-MAC to far end BEB B-MAC from this response and populates the I-component table accordingly, whereupon subsequent traffic between that C-MAC pair uses unicast communication over the backbone.

The complete encapsulation provides for a comprehensive customer–provider demarcation point. The service provider network only transports frames in a provider frame format containing provider administered identifiers. This allows the service provider to separate the topologies used by different customers, or aggregations of customers, by controlling the mapping of I-SIDs to different B-VLANs. Many customers can be supported on a single B-VLAN. It also isolates the behavior of incompetent or malicious customers from the core of the network.

This service identifier thus allows for a greater degree of flexibility in managing services than hitherto, by allowing their complete independence from the topology.

The other advantage of encapsulation is that customer addresses and customer MAC learning are isolated to the provider edge, with the adaptation function providing the mapping between the customer MAC space and the provider MAC space. As the number of BEBs is orders of magnitude lower than the number of customer MAC endpoints supported by the PBBN, the overall scalability of bridging increases by a corresponding amount. Scalability can now be global as interconnected sets of C-MAC addresses are held only at the edge of the network, and moreover, only at those BEBs which have registered an interest in the specific service.

---

[1] This is where SPB deviates slightly from PBB. Because PBB is based on spanning tree, the forwarding tree is common to all BEBs, and PBB can use a common multicast address for an I-SID that is used by every source hosting an instance of that I-SID. SPBM is required to use a unique multicast address per source per I-SID since a unique MAC-based tree per source is needed as a consequence of shortest path forwarding.

Furthermore, this encapsulation has the merit that operations, administration, and maintenance (OAM) procedures are significantly simplified, as the provider edge can now be instrumented independent of the customer addresses. Finally, this separation allows the control plane functions of the carrier to be completely independent of the customer, and vice versa. In particular, there is no need for the carrier to peer with the control plane of all of his customers; the carrier is just providing a completely isolated multipoint-to-multipoint (mp2mp) LAN segment to the customer, and the customer may run over that what he chooses.

SPBM treats unicast B-MAC addresses as falling into two classes as far as backbone control plane operation is concerned. These are

- Nodal MACs,
- Port MACs.

Nodal B-MACs are the SPBM equivalent of the "loopback address" of an Internet Protocol (IP) router, and are the way in which the IS-IS instance on an SPBM node address their peers on other nodes. They may also be used for addressing user data frames if the receiving node can determine how to process the frame from the I-SID only.

However, the PBB forwarding model which SPBM inherited allows multiple granularities of B-MAC addressing in the forwarding path:

- at the nodal level, as above,
- also at the card level, the processing subsystem level, and the customer backbone port (CBP) level.

The reason for this is to allow implementations to be optimized:

- nodal level addressing allows the greatest scalability, but does require per I-SID virtual switches to be implemented on the node NNIs,
- the more granular addressing options allow the NNIs of a node to identify the target virtual switch on the basis of B-MAC alone.

There are two important points to make about these two address classes:

1. Port B-MACs play no part in topology determination or path calculation, which use only the nodal addresses. Port B-MACs are associated with their nodal B-MAC only at the time at which forwarding tables are determined.

2. All SPBM nodes automatically install nodal B-MACs for all other nodes on all B-VID planes. This provides fully connected internode connectivity at all times by default, for example, for use by OAM. To enhance scalability, port B-MACs are installed using the same criteria as per service multicast forwarding; in other words, a node only installs the port B-MACs associated with a service if it lies on the shortest path between two nodes which host endpoints of that service.

All unicast B-MAC addresses and I-SIDs are known to and distributed by IS-IS. Instead of distributing the multicast addresses in SPBM, they are constructed locally by creating a unique source-specific multicast address according to a "well-known" algorithm. A 20-bit identifier called the shortest path source ID (SPSourceID) identifies the source bridge for multicast forwarding. A Group MAC address is formed by concatenating the SPSourceID with the 24-bit I-SID value. The SPSourceID is unique in the network and therefore confers uniqueness on the algorithmically constructed multicast address.

Ethernet's support of up to 4094 VLANs permits multiple sets of equal cost trees (ECTs) to be implemented for both SPBV and SPBM in order to support multipath forwarding over multiple fully connected planes. An "SPT set" corresponds to an individual instance of a single plane fully connecting the network. For SPBV, multiple SPVIDs are used in the construction of each SPT set. For SPBM, an SPT set is delineated by a B-VID. Multipath is only really useful if there is some degree of path diversity between SPT sets, hence an SPT set is typically associated with a particular ECT algorithm for path generation.

The final aspect of both the SPBV and SPBM data planes is the data plane OAM. Ethernet OAM (the IEEE 802.1ag and Y.1731 tool suites) all operate entirely in the data plane, because in bridged Ethernet the routing system is "flood and learn," and there is no control plane. Since SPB makes no changes to the Ethernet data path or the semantics of the data plane identifiers, the entire OAM tool suite can be inherited unaltered.

SIDEBAR: *The Metro Ethernet Forum (MEF) Service Models and Interfaces*

The basic MEF service set describes three connectivity models in the specification MEF 6.1. These are E-LINE, E-LAN, and E-TREE as well as their virtualized (or tagged) equivalents:

- E-LINE corresponds to a p2p Ethernet tunnel,
- E-LAN is an mp2mp LAN segment.
- E-TREE is a split horizon client–server variation on the LAN segment model, which is particularly useful for backhaul and content distribution applications. In an E-TREE, leaves can only communicate with roots, while roots can communicate with both leaves and other roots.

The MEF definitions go significantly further than simply discussing the connectivity primitives, casting these definitions specifically in terms IEEE 802.1ad, and then continuing further to define an entire service architecture including interfaces and instrumentation around these definitions. We have found it useful to consider the 802.1ah and 802.1aq I-SID instantiations of these services to be exact matches of their 802.1ad equivalents, and we can extend the umbrella of these terms directly into the SPB space. The MEF, however, has not defined interfaces in terms other than that of 802.1ad.

The IEEE has been studious in ensuring backward compatibility during every step of Ethernet's journey. The PBB (and therefore SPBM) network model maps the IEEE 802.1ad S-tagged service to the I-SID while preserving all of the attributes of connectivity.

SPB TECHNOLOGY: THE CONTROL PLANE

The IS-IS Routing System Requires Modest Enhancements

The IS-IS link state routing system is put to work to control SPB configuration. IS-IS is uniquely suited to this task due to its robust, standards compliant implementation and many years of live deployment.

IS-IS, the base protocol used by SPB, is commonly associated with IP, but is in fact not dependent on IP at all. IS-IS is a pure Layer 2 protocol, and is capable of discovering a network Layer 2 topology through the use of both a Hello protocol to its immediate neighbors, and by using a flooding protocol (link state updates) to all other nodes in the network. The Hello protocol is used to learn the identifiers (MAC addresses) of the nodes immediately adjacent to a node. The flooding protocol is used to advertise information throughout the IS-IS domain, about a node's immediate neighbors and, in the case of SPBM, about attached service endpoints (I-components corresponding to E-LINE, E-LAN, and E-TREE service instances).

IS-IS in effect provides the distributed database upon which each SPB node executes computations. SPB can therefore be thought of as a sophisticated computation that takes network topology and service information endpoint provided by IS-IS as input, and produces FDBs as an output.

The key factor that allows the collapse of all requisite functionality into a single control protocol is that the Ethernet data plane is fully self-describing, and Ethernet frames transit the network unmodified. The importance of this cannot be overemphasized. The addressing and service identifiers are globally unique network-wide. This property eliminates the need for signaling or any form of per node personalization of the data as an additional convergence step, which is a major advance. Signaling only becomes necessary when forwarding state is locally unique, since local-to-local relationships (such as label switching) must be signaled along every path. By contrast, with SPBM's globally unique MAC/VID addresses, any topology change flooded as a single database update provides all nodes in the network with sufficient information to compute the new network configuration.

The elimination of signaling and the integration of service knowledge into a single control plane radically simplifies the control structure, collapses the number of steps to network convergence, and eliminates race conditions between control protocols.

The next sections introduce the information model used by SPB and summarize the extensions to IS-IS required.

Visual Model of Control Plane Information

These next sections introduce the new information items needed for IS-IS for SPB.

The new items associated with a node are modest in number. Referring to the figure above, the nodal nickname, known formally as

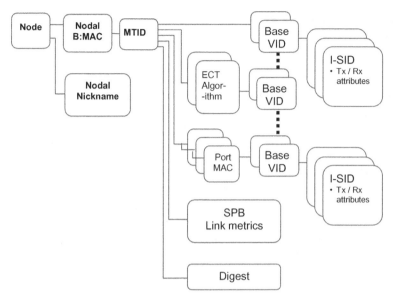

Figure 1.4   New information items in IS-IS for SPBM.

the SPSourceID, is the 20-bit value used to construct the service-specific (S,G) multicast addresses. SPB has its "own" link metric field, to avoid any interaction with other IS-IS applications. The digest is a compact topology summary, used to determine whether or not nodes share an identical topology view, which is a key part of the synchronization process used to guarantee loop-free forwarding at all times.

The remaining items are the nodal unicast MAC addresses, B-VIDs, and the services associated with each. The nodal B-MAC is the SPBM equivalent of the loopback address, and need be the only externally visible address in an SPBM domain. However, PBB permitted different granularities of B-MACs, to allow implementation trade-offs to be made. Multiple B-VIDs for load spreading and traffic engineering may be associated with SPBM operation, and hence the services (I-SIDs) and port B-MACs may only be associated with a single B-VID.

The information model for SPBV is a significantly simpler subset of the one for SPBM. It is presented in Chapter 3, in the more detailed treatment of the control plane ("Visual Model of Control Plane Information," p. 89).

Link State Packet (LSP) Extensions for Link State Bridging

Link state bridging introduces no new protocol data units (PDUs) to IS-IS and adds only new type-length value (TLV) fields and sub-TLVs to the existing IS-IS PDUs. These are briefly and informally enumerated now, with fuller details of each type and its parameters deferred until later ("New IS-IS TLVs for Link State Bridging," p. 90):

(a) *The Multitopology Aware Port Capability (MT-PORT-CAP) TLV.* It differentiates topology instances in Hello PDUs. Each IS-IS topology allows only one metric per link; multitopology (MT) allows the use of different IS-IS metric sets running on the same topology if this is desired for manipulation of preferred traffic paths.

   • this carries an MT identifier (for possible use in future), and

   • an overload bit specifically for use by link state bridging, used to indicate whether the bridge can be used for transit, following the analogy of the generic IS-IS overload bit.

(b) *SPB MCID Sub-TLV.* This sub-TLV is added to an IS-IS Hello (IIH) PDU to communicate the multiple spanning tree configuration identifier (MCID) for a bridge. This digest is used to determine when adjacent bridge configurations are synchronized. The data used to generate the MCID is the allocation of VIDs to the various protocols used by the bridge, which is populated by configuration, and the digest is based on a cryptographic hash of these allocations. Adjacent SPB bridges may only use the link between them for SPB traffic if their digests are identical. Two MCIDs are carried to allow transitions between different but nonconflicting configurations.

   The important information elements are:

   • The MCID and the auxiliary MCID. The complete MCID identifies an SPT region, and its computation is defined in [SPB].

(c) *SPB Digest Sub-TLV.* This TLV is added to an IIH PDU to indicate the current topology digest value. Matching digests indicate (with extremely high probability) that the topology view between two bridges is synchronized, and this is used to control the updating of multicast forwarding information. Digest construction is considered later in the control plane description, under "Agreement Digest Construction details" (p. 115).

During the propagation of LSPs the Agreement Digest may vary between neighbors until the key topology information in the LSPs is synchronized. The digest is therefore a summarized means of determining agreement on database consistency between nodes, and may hence be used to infer that the nodes agree on the distance to all multicast roots. The SPB Digest sub-TLV contains the following key information:

- A (2 bits) The *Agreement Number* 0–3 (a rolling count sequence number), which aligns with the *Agreement Number* concept fully described in [SPB], used to guard against control packet loss.
- D (2 bits) The *Discarded Agreement Number* 0–3 which aligns with the *Agreement Number* concept of [SPB].
- Agreement Digest. This digest is used to determine when IS-IS is synchronized between neighbors, and comprises a hash computed over the set of all SPB adjacencies (all edges) in all SPB MT instances. This reflects the fact that all SPB nodes in a region must have identical VID allocations, and so all SPB MT instances will contain the same set of nodes.

(d) *The Multitopology Aware Capability TLV.* It differentiates topology instances for other SPB TLVs.

(e) *SPB Base VLAN-Identifiers Sub-TLV.* This sub-TLV is added to an IIH PDU to indicate the mappings between ECT algorithms and Base VIDs. This information should be the same on all bridges. Discrepancies between neighbors with respect to this sub-TLV are temporarily allowed during upgrades (e.g., during the assignment of new ECT algorithms to Base VIDs), but all active Base VIDs, as declared by the state of the Use-flag below, must agree and use the same ECT-ALGORITHM.

The key information element is a list of ECT-VID tuples, each comprising

- The ECT-ALGORITHM (4 bytes), which declares that the advertised algorithm is being used on the associated Base VID.

- The Base VID that is associated with the SPT Set defined by the ECT-ALGORITHM which supports a single VLAN over the SPT region.
- A Use-flag, which is set if this bridge, or any bridge in the SPB region, is currently using this ECT-ALGORITHM and Base VID. This is formed from the logical OR of the U-bits (found in the *SPB Instance Sub-TLV* below), and is used to ensure orderly upgrade when new Base VIDs are introduced.

(f) *SPB Instance Sub-TLV.* The SPB Instance sub-TLV announces the SPSourceID for this node/topology instance. This is the 20-bit value used for formation of multicast DA addresses for frames originating from this node/instance. The SPSourceID occupies the upper 20 bits of the multicast DA together with 4 other bits (see the SPBM multicast DA address format). This sub-TLV is carried within the MT-Capability TLV in the fragment ZERO LSP.

This TLV carries several information elements used for compatibility with bridges running STP, and these are not enumerated here. The important information elements from the point of view of SPB comprise the following:

- Bridge Priority is a 16-bit value which together with the low 6 bytes of the System ID form the spanning tree compatible Bridge Identifier. This Bridge Identifier is a unique value which is used in SPB by the ECT tie-breaking algorithms.
- The SPSourceID is a 20-bit value used to construct multicast DAs for SPBM multicast frames originating from the node which originated the LSP containing this TLV.
- A list of ECT-VID tuples. Each ECT-VID tuple defines one VLAN, and comprises the ECT-ALGORITHM and Base VID information given earlier, under the SPB Base VLAN-Identifiers sub-TLV, and adds a declaration of whether any I-SIDs are assigned to this Base VID at this node (the U-bit). Each ECT-VID tuple also declares the SPVID used by this bridge to identify it as the root of a shortest path tree when operating in SPBV mode.

(g) *SPB Instance Opaque ECT-ALGORITHM Sub-TLV.*

(h) *SPB Adjacency Opaque ECT-ALGORITHM Sub-TLV.* There are multiple ECT algorithms already defined for SPB; however, additional algorithms may be defined in the future. These algorithms will use this optional TLV to define new algorithm tie-breaking data. There are two broad classes of algorithm, one which uses nodal data to break ties and one which uses link data to break ties, and so as a result two identically formatted TLVs are defined to associate opaque data with either a node or an adjacency.

(i) *SPB Link Metric Sub-TLV.* The SPB Link Metric sub-TLV occurs within the Extended Reachability TLV or the MT Intermediate System TLV.

The important information elements are

- SPB-LINK-METRIC indicates the administrative cost or weight of using this link as a 24-bit unsigned number. Smaller numbers indicate lower weights and are more likely to carry traffic. Only one metric is allowed per topology instance per link.

- Sub-TLVs can include an opaque ECT Data sub-TLV, whose first 32 bits are the ECT-ALGORITHM to which this data applies. This sub-TLV carries opaque data for the purpose of extending ECT behavior in the future.

(j) *SPBM Service Identifier and Unicast Address Sub-TLV.* The SPBM service identifier and unicast address sub-TLV is used to declare service group membership on the originating node and/or to advertise an additional (port) B-MAC address by which the I-components supporting the declared service instances may be reached. The SPBM service identifier sub-TLV is carried within the MT capability TLV.

The information elements are

- A single B-MAC address, which is a unicast address of this node. It may be either the nodal address, or it may address a port or any other level of granularity relative to the node.

- The Base VID (and hence the ECT-ALGORITHM) to which the following list of service identifiers are assigned.

- A list of service identifiers: ISID #1 . . . #N are 24-bit service group membership identifiers. Each I-SID has a transmit (T) and receive (R) bit which indicates if the membership is as a transmitter/receiver or both (with both bits set). In the case where the transmit (T) and receive (R) bits are both zero, the I-SID is ignored for the purposes of multicast computation, but the unicast B-MAC address must be processed. In this scenario, edge based replication of broadcast, multicast, and unknown frames replaces the use of an (S,G) multicast distribution tree.

The SPBM service identifier sub-TLV is carried within the MT capability TLV and can occur multiple times in any LSP fragment.

(k) *SPBV MAC Address Sub-TLV.* The SPBV MAC address (SPBV-MAC-ADDR) sub-TLV is not used by SPBM, only by SPBV. It contains the following information elements:

- SR bits (2 bits), which derive from Multiple MAC Registration Protocol [MMRP], specifying (typically at port level), the service requirements external to the SPBV region applicable to the following set of group addresses.

- SPVID (12 bits); the SPVID and, by association, the Base VID and the ECT-ALGORITHM and SPT set that the MAC addresses defined below will use.

- A list of Group MAC addresses which declare this bridge as part of the multicast interest for these addresses, with bits to indicate if membership is as a transmitter, a receiver, or both. Pruned multicast trees can be constructed by populating FDB entries for the subset of the shortest path tree(s) that connect the bridges supporting that MAC address as a receiver. This replaces the function of the 802.1ak [MMRP] for SPTs within an SPBV network, and allows the semantics of MMRP messages received at the edge of an SPBV region to be flooded across it.

SPB TECHNOLOGY: PATH COMPUTATION

To this point, the SPB data path and the extensions to IS-IS needed to support them have been the main focus. However, several other issues still needed to be resolved, largely related to the challenge of how to provide a node with the data it needs to compute its forwarding state. The following section also highlights how SPB computes forwarding state for only a subset of all destinations so that the familiar E-LINE, E-LAN, and E-TREE services may be rapidly created in very large quantities.

Computing Forwarding State

SPB nodes learn the topology of the network in a standard IS-IS link state fashion, and once each node has learned the topology then the shortest paths for unicast and multicast traffic are determined by simple shortest path (Dijkstra) computations against the data distributed by IS-IS.

SPBV scopes multicast (for an entire VLAN) using SPVIDs, while SPBM scopes each multicast receiver set (per source per I-SID) using service-specific multicast addresses, each within a single B-VID if multiple paths are being used. The decision process for computing multicast SPB forwarding state for both modes of operation is fairly straightforwardly described. Every node asks itself a simple question: "Am I on the unique shortest path between a given pair of nodes and do those nodes participate in a common service?" This requires that some variation of the "all-pairs shortest path" algorithm is run against the link state database. When an SPBV node finds itself on the shortest path between two bridges for a given VLAN it installs those bridges' SPVIDs associated with that VLAN on the appropriate ports in its FDB.

When an SPBM node finds itself on the shortest path between any two BEBs for a given SPT set/B-VID, it checks the transmit/receive attributes of I-SIDs assigned to that B-VID on those BEBs, and if it is on the shortest path between a transmitter and a receiver on an I-SID, the node installs any associated unicast port MACs and locally constructed multicast MAC addresses in the local FDB. In this manner, only bridges involved in the forwarding of traffic for a service will ever install forwarding state for that service. When all nodes in a given path have completed the computation and installed forwarding state, a given path will be complete end-to-end.

The routing system is "single touch" for service addition and removal; only the node that is joining or leaving needs to be configured with the service change. All other nodes will be informed by flooding in IS-IS, and the multicast trees and unicast forwarding paths will be adjusted accordingly to keep the routing optimal. It is an important property of shortest path trees that neither addition nor removal of a node alters the tree to other nodes, so changes to service membership do not disrupt unaffected nodes.

Per-Service-Instance Routing and Forwarding

Computation of IP forwarding tables traditionally requires only a single shortest path calculation, with the computing node placed at the root, to determine the next hop to the set of destinations. SPB requires a node to compute whether it has a transit role for traffic between all possible pairs of nodes in the network, and therefore requires the computation of "all-pairs shortest paths." Although this is computationally intensive, some two orders of magnitude more performance is now available in embedded processors compared to when shortest path calculation was first deployed. Furthermore, the modern trend in processor architecture is to move to multiple cores, and the $N \times$ Dijkstra calculations to perform "all-pairs shortest paths" for an N node network naturally partition into coarsely parallel threads on as many cores as are available.

With its use of the SPVID as a network-wide source node identifier in the data plane, SPBV builds a complete mesh of broadcast trees, one per node per SPT set.

SPBV has the same limitations as QinQ in that the SPVID set is overloaded to be both a topology and a service instance. An immediate consequence is that the scalability of services is drastically constrained due to the VID consumption needed to construct basic connectivity. SPVIDs are unidirectional and this does permit the construction of MEF services in an SPBV network. The Base VID defines the service association, so that E-LINE and E-LAN services each map to a single Base VID that will have two or more SPVIDs associated with it, equal to the number of Provider Bridges participating in the VLAN. E-TREE requires the use of two Base VIDs, both associated with the same ECT algorithm such that congruence is preserved in order for shared VLAN learning to work. One Base VID defines the set of leaf to root paths, and the other Base VID the set of root to leaf paths.

SPBM can exploit the "all-pairs shortest paths" computation more fully, by building per service multicast trees which are each a strict subset of the multicast tree rooted on the node hosting an instance of the service.

After SPBM nodes complete their "all-pairs shortest paths" calculation, if two nodes require just a simple E-LINE service, the computation will result in the installation forwarding state on all nodes between the two nodes on the shortest path. Essentially, SPBM will create a p2p connection for that service.

In SPBM, if a third member (node) is now added to the service, transforming it from an E-LINE into an E-LAN, SPBM will automatically compute and create forwarding state for this service instance from each member to the other two members along shortest paths.

The E-TREE service deserves special discussion because SPBM solves this in a very simple manner. When a node advertises that it has a member of a service instance, it indicates whether that member will be a transmit-only, a receive-only, or a transmit–receive member of that service. This allows "split-horizon" behaviors to be created.

An SPBM E-TREE service instance is therefore formed by using two I-SIDs in a direct analogy to QinQ's asymmetric VID. On one I-SID, transmit-only "spoke" members can send only to "hub" members which are receive-only. On the other I-SID, "hub" members are receive + transmit (so that they can communicate with each other as well as with spokes), and all "spoke" members are receive-only. The SPBM computation algorithm takes the transmit/receive attribute into consideration and uses it to create the unidirectional state as required between the members, thus ensuring that they cannot violate transmit- or receive-only constraints placed on them by the operator, and that bandwidth consumption is minimized. Very simply, when a node finds itself on the shortest path between two nodes with an I-SID in common it will check the transmit/receive attributes for both ends for each direction. If there is a transmit attribute at one end and a receive attribute at the other, the destination multicast MAC for the transmitter is constructed and installed in the FDB. At the edge of the network, the I-components perform the functional equivalent of "shared VLAN learning" for the mapping customer MAC addresses to backbone MAC addresses in order for the bridging aspects to operate properly.

A topic of concern in the industry is the notion that frequently a BEB or MPLS/VPLS-PE will host a root instance and separate and distinct leaf instance(s) for a given E-TREE. This can be a frequent occurrence in networks with a significant backhaul component which also has service grooming capability. This is actually a more subtle problem than the description above would indicate, as the choice to colocate the root and leaf instantiation on a common subsystem is both an implementation issue and/or an operational practice problem. When this scenario occurs for SPBM, the virtual switch instance (VSI) hosting the combined root and leaf functionality will simply advertise transmit and receive interest in both I-SIDs. The internal implementation will determine the steering of frames to and from the root and leaf ports on the basis of the I-SID received for egress traffic, or inferred from port/ VID attributes for ingress traffic, and the rest of the SPBM network will simply fill in the proper connectivity to the other roots and leaves. Where the node implements the VSIs in different subsystems, and the implementation cannot "hide" that fact, the node will be obliged to advertise different B-MACs (identified as "port MACs" earlier) in order to correctly steer frames within the BEB.

A consequence of the generalized provisioning and fulfillment model is that SPBM allows single-point or one-touch provisioning— the "Holy Grail" so to speak of multicast/E-LAN service solutions. An operator may add a new member to an E-LAN service by configuring only that new member, without disruption to the existing members and with a fulfillment time corresponding roughly to the network conver- gence time. The distributed nature of SPBM computations and the piggybacking of service information in the same protocol as is used to distribute topology mean that SPBM can perform all of its functions without additional provisioning or protocols.

There are of course numerous other permutations possible with these service attributes. For example, a unidirectional E-LINE is just two nodes, where one advertises transmit-only membership while the other advertises receive-only membership. Likewise, one can create more elaborate communities with, say, two transmitters and multiple receivers to allow for redundant head-end transmitters. The mechanism employed by SPBM to form service instances is simultaneously elegant, simple and powerful, and highly scalable. This is fundamentally because a calculation is far simpler and faster to perform than signaling or other message-based solutions to create an individual service instance.

How Symmetry and Congruence Are Preserved

The challenges of using a simple shortest path computation involve ensuring that both unicast and multicast traffic are kept on the same shortest path, and that the chosen path is the same in the forward and reverse directions for both unicast and multicast traffic, even when there are multiple equal cost candidate paths available. In essence, the connectivity across a network between any two points in a converged network should behave like a bidirectional p2p link.

This absolute symmetry (which is an inherent and desirable property of STP) is very important because without it, many of the Ethernet OA&M mechanisms lose their value. Further, the overlaying of client bridging on this infrastructure avoids misordering of frames and race conditions between unicast and flooded unknowns. If traffic is spread around, this would no longer be true. There are other important benefits to this symmetry and determinism, discussed below.

The solution to this challenge is to use a shortest path tree not only for the unicast routes, but also for the multicast routing from a given source. The end result is that each node in the network has its own source tree, which originates from itself and reaches all other nodes. When equal paths on any tree are resolved by the deterministic tie-breaking algorithm described below, all nodes will then choose the identical source tree rooted on any node. Now, instead of knowing about just one spanning tree, as is the case with the existing Ethernet STP, in SPB each node knows about one shortest path tree from every node in the network (in the multicast world, this type of per-source tree is referred to as an (S,G) tree, whereas a single tree for all members of the community (e.g., as constructed by the STP protocol) is referred to as a (*,G) tree. It should be noted in the case of a (*,G) tree that Ethernet split horizon forwarding ensures a sender does not get a copy of a multicast frame it has sent to the group looped back to it, a property which is essential for the prevention of loops. IEEE 802.1aq preserves this property.

Using these trees, every transit node in the network can easily forward unicast traffic along the shortest path simply by hop-by-hop destination lookup, and every node can multicast or broadcast traffic along the same route as the unicast traffic as long as it knows which node originated the multicast or broadcast frame.

Tiebreaking

Frequently, the shortest path between a source and destination in a network is not unique. There may, in fact, be dozens of equivalent shortest paths between a source and destination. SPB requires that for a given SPT set, every node agrees upon the same one of these paths and that determination must be made by each node without reference to other nodes, because no signaling is used. This requirement is met by a symmetric tie-breaking algorithm which, when executed by every node in the network whenever offered a choice of shortest paths, still results in a network-wide consistent decision as to which end-to-end path is chosen. The determinism of SPB has the added benefit of allowing accurate prediction of exactly where traffic will go, using for example an offline network planning tool.

The requirements on the distributed tie-breaking algorithm can be reduced to needing independence of the computation order, and independence of the network position of the computation. Each bridge has a unique Bridge ID. A path ID is specified as the lexicographically sorted list of the Bridge IDs which the path traverses, which is therefore also guaranteed to be unique. This satisfies the requirements for distributed tiebreaking, because all nodes find the same paths between any two endpoints, and the sorting process relies only on the values of the Bridge IDs in the path, which is not dependent on the order of computation. Thus, all nodes implementing the same logic choose the same path from the available options, for example, the one having the lowest path ID after ordering the Bridge IDs from lowest to highest in the path ID.

This algorithm has a further benefit from a computational perspective. The sorting algorithm results in the property that any segment of a shortest path is also itself a shortest path. As a consequence, as soon as multiple shortest paths forming a segment of a longer path have been resolved, all the state associated with the rejected paths can be discarded because it is known that it will never be required again. This simplifies the computation, since the amount of state to be carried forward as a Dijksta calculation progresses across the network can be continually pruned.

Finally, it is easy to see that there is a dual of the algorithm above, in which the selected path is the one having the highest path ID following highest to lowest ordering of Bridge IDs in the path ID. Although not guaranteed to find a diverse path if one exists, this technique is weighted toward doing so; as such it forms the first example of the techniques for load spreading across multiple equal cost paths which are introduced below and discussed at more length in "Load Spreading: Equal Cost Trees" (p. 119).

Exploiting Multiple Paths in SPB

Shortest path forwarding enables an inherent improvement in utilization of a densely meshed network, because all links can be used, and none need be blocked for loop prevention. It is possible to get even better utilization by allowing the simultaneous use of multiple equal cost shortest paths. The IEEE 802.1aq standard initially supports up to 16 different shortest paths between any pair of endpoints and provides an extension mechanism to permit future enhancements to be supported. This is achieved by manipulation of the tie-breaking criterion used to select between multiple equal cost shortest paths.

The previous section on Tiebreaking (p. 27) showed that consistent distributed path computation can only be achieved if all bridges make the same path choices, and a deterministic algorithm to do this was introduced. This can be extended further, and by using a set of globally defined transformations of the Bridge ID prior to sorting to form the path ID for each forwarding plane, different path sets are selected and each set is associated with a single VLAN (forwarding plane), and hence a set of SPVIDs for SPBV, and a single B-VID identical to the Base VID for SPBM. IEEE 802.1aq specifies 16 Bridge ID transformation methods, using the XOR of the Bridge ID with a "well-known" set of masks, and in addition makes possible the definition and application of further tie-breaking methods in the future.

The ECTs have unique attributes. First, the path congruence property means that IEEE 802.1aq actually supports equal cost routing for multicast and broadcast traffic as well as unicast. Second, since these are **end-to-end** paths, and not a hop-by-hop spreading function; assignment of traffic to a path is done once, at the ingress to the network, and so the operator can avoid random assignment, and instead place the traffic based on estimates of loading. This can be viewed as a low-overhead type of traffic engineering, in which services are not individually microengineered, but are assigned to a specific forwarding plane.

SPB TECHNOLOGY: LOOP AVOIDANCE

The SPB Approach: How It Works

Moving from spanning tree to a distributed routing system and mesh connectivity enables a mechanism to address transient loops at a much finer granularity than port blocking. For unicast forwarding, a routing loop is at best an inconvenience and at worst a chronic drain on bandwidth. For multicast forwarding, looping can be catastrophic, especially if a loop feeds back into another loop, resulting in an exponential increase in the bandwidth consumed in the network, in turn causing nearly instantaneous network "meltdown."

The combination of bidirectional symmetry and unicast/multicast path congruency between any two SPB nodes means that the FDB already has sufficient information to suppress most loops. In a stable and converged SPBV network, a frame from a given root can arrive only at ports which are on the shortest path to that root, which are also the only valid ports of receipt for the SPVID associated with the root. Similarly for SPBM, a frame **from** a given MAC address should only arrive at a port which is also on the shortest path **to** that same MAC address. A frame from a given root arriving at an unexpected port is an indication of a problem and potentially a loop. For SPBV, port membership of the VLAN (i.e., SPVID) in question determines whether the frame is valid. SPBM requires a simple modification to Ethernet source learning to convert it into a "reverse path forwarding check" (RPFC). RPFC simply checks the port of arrival of a frame against the expected egress port for that frame's source MAC address; frames arriving on an unexpected interface are silently discarded.

It can be shown that although the addition of VID or MAC-based RPFC substantially improves protection against loops in a routed environment, it cannot guarantee to eliminate a loop under some pathological conditions. It is nonetheless worth pointing out that even if such a loop does form, RPFC ensures that it must be "closed," preventing any new frames entering the loop, and thus guaranteeing that an exponential increase in the number of looping frames cannot occur. This is because RPFC allows any node to receive frames from only a single ingress port, so flows from two or more directions can never merge.

Nonetheless, it was desired that a technique offering absolute assurance against loop formation should be available in SPB. Such a technique has been specified, which uses the IS-IS topology database at a bridge to identify potentially hazardous changes, and to trigger topology database synchronization with neighbors before forwarding state associated with such hazards is installed.

A high level summary of this approach is that when an SPBV node receives a topology change, it

1. computes its unicast topology and FDB;
2. when it then determines that the shortest path distance to a root node has changed, so that there is a possibility a loop could form, it blocks the SPVID(s) associated with that node on all its ports;
3. it unblocks the SPVID entries related to changed trees (removed at step 2 above) on each port only when it knows that its control plane is synchronized with the neighbor connected to that port.

Similarly for SPBM, when an SPBM node receives a topology change, it

1. computes its unicast topology and FDB;
2. when it then determines that the shortest path distance to the root of a multicast tree has changed, so that it is possible a loop could form, it removes the multicast FDB entries related that to tree at the same time as it installs the updated unicast FDB entries;
3. the node can then install multicast entries for trees which have changed, but which are considered "safe" (because the shortest path distance to the root of that tree is unchanged), without reestablishing topology agreement with its neighbors, as the existing relationship has not changed;
4. the node installs the multicast FDB entries related to changed trees (removed at step 2 above) only when it knows that its control plane and unicast FDB state is based on a view of topology which is synchronized with its neighbors.

It should be noted that this process is less intrusive on the network in SPBM due to its use of per service multicast (rather than broadcast) trees, and due to its ability to distinguish unicast from multicast MAC forwarding, and so to give unicast different treatment.

The test above, for **any** change of distance to the root of the tree, is more aggressive than is strictly necessary, but it still ensures that trees unaffected by a fault carry on forwarding. This test has the major attraction that it is a nodal test, and is trivial to implement, because it is a simple per tree comparison of "before" and "after" distances to the root, and does not require per port calculation.

The optimal test is more complex to implement, but the actual requirements for handshaking with neighbors are still fairly straightforward:

- If a node believes that it has moved closer to a given root, it needs to handshake with its (new) parent node on that tree before installing the affected multicast entries. This ensures that the node's new parent is guaranteed to be closer to the root than the node itself;
- If a node believes that it has moved further from a given root, it needs to handshake with any child nodes on that tree whose last known (synchronized) distance from the root was closer than this node now believes itself to be. Only then may multicast be re-enabled on the port connecting to a child node. This ensures that the children still remain further from the root than this node.

This approach has a number of desirable properties. First, we maintain uninterrupted connectivity for multicast trees unaffected by the topology change, which exploits a key inherent property of link state protocols. Second, synchronization of multicast updates does not need to be ordered from the root; nodes can safely reinstall affected state as soon as they are synchronized with relevant peers, because if a peer has not achieved the required synchronization further up the tree, its own lack of installed multicast state "protects" the downstream nodes. Finally, the delay which synchronization would normally be expected to incur is largely eliminated, as the required handshaking with peer nodes can be done in parallel with the computation of the multicast FDB.

Synchronization with neighbors involves the exchange of a digest of the current IS-IS database, in order that both nodes can agree on the network topology they have used when computing loop-free forwarding paths (by removal when required of multicast state that was at risk of causing a loop). The principle is that if their exchanged digests match exactly, then they must also both agree on the distance to all roots. The digest is constructed in a way that maximizes how authoritative the comparison is, and also minimizes the overall computation. This is possible because the requirement is simply to determine whether two nodes agree on their topology model; if they differ, they differ, and there is no need for the digest to allow differences to be resolved, because a node determines what multicast state should be removed by the differences between its **current** view of the topology, and the topology view it held when it **was** last synchronized.

Summary of Topology Digest Construction

The requirements which must be met by the Topology Agreement Digest are:

- to summarize the key elements of the IS-IS link state database in a manner which has an infinitesimal probability that two nodes with differing databases will generate the same Digest;
- to have a very low incremental computation overhead because in general, link failure and repair are isolated events, and so it is very desirable that a single event should not require complete recomputation.

To achieve this, the Topology Agreement Digest field comprises six elements:

- the Digest Format Identifier
- the Digest Format Capabilities
- the Digest Convention Identifier
- the Digest Convention Capabilities
- the Digest Edge Count
- the Computed Topology Digest

The first four fields are provided to preserve extensibility, allowing development of alternative digests in the future if required, and will not be discussed further here.

The Digest Edge Count is a 2-byte unsigned integer. Its purpose is to offer a summarization which is simple to compute and powerful in detecting many simple topology mismatches. In the light of the use of a strong hash for computation of the Computed Topology Digest, the Edge Count can be seen as a historical hangover from a time when a simpler multiplicative hash was envisaged.

This value is the sum modulo 2^{16} of all edges in the SPB region. Each p2p physical link is counted as two edges, corresponding to its advertisement by IS-IS in an LSP flooded from either end of the link.

The overall procedure for constructing the final component, the Computed Topology Digest, is to:

- form a signature including every edge in the topology by computing the MD5 hash (RFC 1321) of the significant parameters of the edge, as defined below;
- compute the Digest as the arithmetic sum of all edges in the topology.

Although MD5 is widely reported to be cryptographically compromised, this is not relevant in this application because there is no motivation for an attack. What is required is a function exhibiting good avalanche properties such that signatures with potentially very similar input parameters have an infinitesimal probability of collision.

This strategy also allows the Computed Digest to be incrementally computed when the topology changes, by subtracting the signatures of vanished edges from the Digest and adding the signatures of new edges.

The input message to the MD5 hash for each edge is constructed by concatenating the following fields in order:

1. the Bridge Identifiers (Bridge Priority ‖ Bridge SysID) of the two bridges advertising the edge, with the numerically larger Bridge ID first;
2. one 3-tuple for each MTID declared in IS-IS. The 3-tuples are declared in descending order of MTID value, with the largest MTID declared first.

Each 3-tuple is constructed by concatenating the following fields in this order:

- MTID value ‖ Link Metric of higher Bridge ID ‖ Link Metric of lower Bridge ID

If an edge is not present in a topology, its SPB Link Metric is set to zero in that topology.

The value of the Computed Topology Digest is the arithmetic sum of all of the signatures returned by presenting every edge message to MD5, treating each signature as an unsigned 16-byte integer and accumulating into a 20-byte integer. Every physical link is seen as two edges, one advertised in an LSP by each bridge comprising the adjacency, and formally the Computed Topology Digest includes both.

SUMMARY

In summary, SPB applies state-of-the-art link state routing to Ethernet forwarding, with the intent of providing a robust and efficient any-to-any infrastructure. In its SPBM incarnation, this is used to support client-server hierarchy to deliver perfect virtualization of traditional enterprise Ethernet, in which virtual LAN segments defined by the I-SID replace physical facilities dedicated to each LAN.

So far we have discussed how SPB functions as it is currently specified. In the next sections we will explore some of the background behind the journey to this point, and provide some insight into the design decisions described to date.

CHAPTER 2

Why SPB Looks as It Does

Having offered a succinct introduction to what SPB is, we now back up, and begin our more comprehensive discussion, not only covering **what** SPB is, but **why** it is as it is, starting with revisiting the network requirements in some detail.

We then embark on a top-down discussion of the SPB technology, starting with its antecedents, which have defined the key attributes of bridged Ethernet (see "History," p. 52), and explaining why these attributes are preserved even though they are no longer strictly necessary for a routed technology in which media access control (MAC) learning is not used (see "Lynchpins: Constraints We Chose to Respect," p. 60). We then discuss how Ethernet concepts, such as the VLAN, are subtly reinterpreted in novel ways, while retaining consistency with their formal specification. We then cover a number of other system aspects of SPB.

We then focus on the SPB control plane, starting at p. 74, under three major headings:

- an introduction to the IS-IS routing system, and the use made of it in SPB. This includes an extensive treatment of multicast in SPB, because this is such a core aspect of Ethernet functionality;

802.1aq Shortest Path Bridging Design and Evolution: The Architect's Perspective, First Edition. David Allan and Nigel Bragg.

- the information model used by IS-IS for SPB, including the new (sub)-TLVs being defined for its use;
- we then devote a significant section to various algorithmic aspects, which cover how loop avoidance is guaranteed, how load spreading is achieved safely, and finally, various insights on techniques for fast computation of the required trees.

THE PROBLEM SPACE

Characteristics of Different Network Regions

Enterprise LAN is the traditional Ethernet domain. However, the objective of the original pre-standard design (known as PLSB) was quite specifically to **widen** the range of applications accessible to Ethernet, not just serve the traditional ones better. These other applications are largely in provider space, have significantly different requirements, and are the subject of the sections immediately following the LAN summary below. This section amplifies on these requirements and extends the discussion on their implications, to provide the background on the motivation for developing SPB.

The Enterprise LAN

Historically, Ethernet was developed to serve enterprise campus networks, and their requirements have shaped Ethernet's evolution until recently:

- such networks are typically hierarchical trees, with multiple layers of aggregation from workgroup switches to campus core;
- for enterprises, communications is a cost and not a core competency, hence something to be controlled, not a profit center. A key aspect of cost containment is "plug and play" networking, with minimal administration;
- for the same reason, while bandwidth is never free, in the enterprise LAN bandwidth is usually cheaper than the recurring operational complexities needed to use it efficiently. As a consequence, sophisticated traffic management has not been widely deployed;

- in an enterprise, network users are typically inside the "ring of trust," and run a common IP address plan, and so there is not the requirement for rigorous virtualization and customer isolation found in carrier networks, where shared infrastructure carries the traffic of many different customers. So, although virtual LANs were developed in Ethernet [802.1Q], their motivation was as much about control of the size of broadcast domains as it was about isolation between groups of users.

The Metro Network

With the increasing dominance of packet traffic in carrier networks, and the emergence of Ethernet as the future ubiquitous physical layer for packet services, carriers have deployed Ethernet services in the metro region using a number of platforms. Initially, Ethernet services were carried using adaptation into point-to-point (p2p) time division multiplexed (TDM) transport technologies such as SONET or SDH, but now Ethernet services are usually deployed on platforms for which packet switching is native.

Calling all of these "Ethernet services" can be confusing, and the emergence of the industry term "connection oriented Ethernet" has added to the confusion. In reality, there are two or three broad classes of packet services emerging:

- there are Ethernet connectivity services, p2p or multipoint, typically sold to enterprise customers for their private intersite communications, and where the service sold is the Ethernet connectivity itself;

- there are many applications where the service actually runs at a higher layer, usually IP, and Ethernet provides a transport and layer 2 presentation for this service. Examples are backhaul from DSLAMs and passive optical network OLTs to the Broadband Network Gateway (BNG) for residential Internet access, and also backhaul from enterprise sites to an MPLS Provider Edge (PE) that delivers [IP-VPN] services;

- there are regulatory environments where p2p backhaul is provided by an access provider, and aggregated traffic is then handed off to one of a number of Internet service providers (ISPs), which adds a wholesale component to the mix.

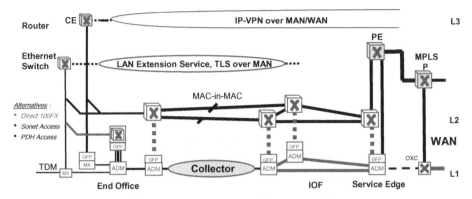

Figure 2.1   Metro network architecture: layer view.

This focus on backhaul results in a metro network architecture which is represented in the figure above. In this, the customer devices at the network edge are on the far left, and the PEs delivering the IP services are on the right, at the edge of the core wide-area network. The function of the lower layers between is to provide transparent bit transport between the IP nodes; traditionally, this was done at layer 1, by SONET/SDH. Now this is increasingly performed by packet transport at layer 2, an Ethernet model being shown above. The key characteristics of this deployment model are

- it is fundamentally a "hub and spoke" architecture, with geo-redundant hubs often deployed for resiliency (e.g., spared BNG/ BRAS sites and voice servers, in a residential context). In this model, the spoke structures may perform traffic aggregation, but the intelligence underlying the service is delivered from the hub;

- in such networks, there is usually limited physical diversity as a consequence of the existing fiber runs and rights of way. If a particular node in the backhaul network warrants protection, a second fiber must typically be installed and/or lit, and so if protection is provided then a single alternate route is all that is likely to be available;

- a further consequence of this architecture is that traffic engineering (TE) and budgeting for resiliency is comparatively straightforward. Traffic at the edge is not substantially aggregated and is sparsely connected, and hence the options to mitigate bandwidth exhaust are limited. Traffic flows from edge nodes to defined hubs in known aggregate volumes. There are still issues to consider, such as the extra oversubscription which is acceptable in the event of route failure, but complex traffic engineering can rarely do much to help; if a route is down, there is typically only one route left;

- in metro networks, the important relationship from the customer's perspective is that between his "appliance" and the IP service function of the provider. The backhaul network must deliver provider-managed service-unaware connectivity, no more, no less. This managed underlay therefore aggregates by geography, and has a major cross-connection function at hubs, where individual connections aggregated by geography are distributed between different services. This is a particular issue in some regulated environments, where the access provider is obliged to hand traffic off to the service provider of the customer's choice.

A major debate in the industry at the present time concerns how deep the service edge function (the Provider Edge node or PE in the diagram above) penetrates into the metro, although the discussion is not often explicitly in these terms. In the past, service edge functions have been highly consolidated into modest numbers of locations because that makes operation and management of the IP/MPLS core network tractable. The concomitant requirement for scalable backhaul was satisfied by SONET/SDH TDM technologies.

This same "consolidated service edge" model is supported by Provider Backbone Bridge-Traffic Engineering (PBB-TE) in the backhaul because it basically translates the SONET operational paradigm to an efficient packet embodiment. However, there are proponents of an alternative "all services, every box" model, in which the IP/MPLS service edge is pushed deep into the metro. This reduces the scope of backhaul network to little more than an access function, but makes the service architecture and its management much more diffuse, and at the same time stresses the core network scaling. It similarly undermines the law of "large numbers" from the perspective of the core network, as the traffic on interfaces near the edge is much less highly aggregated, and as a consequence needs higher bandwidth dilation to offer a similar level of service to a network with a much deeper backhaul component.

A related though distinct trend is advocacy of relatively dynamic service placement, with the intent of achieving optimal load balancing across service edges. This has been used in limited domains before, for example, in the load-balancing and layer 3 resiliency model supported by multiple BRAS sites.

This constrained topology can be supported by straightforward E-LAN connectivity at layer 2. However, its extension to more general scenarios will require more flexibility at layer 2 to support less restricted service migration, and the desirability of a layer 2 control plane will become more apparent.

SIDEBAR: *Ring versus Mesh*

Traditional transport network technologies have typically been intended for deployment as rings. The major production data network technologies are all based on meshes. Especially in the metro, both models coexist, and this has been a source of some confusion, and it is worth briefly summarizing the issues.

Physical fiber is usually deployed in ducted rings. There are three main reasons:

1. this maximizes the number of sites passed per unit length of ducting (trenching is expensive), and so is the lowest cost way of providing network access;

2. it provides a straightforward resiliency model, protecting against any single physical cut;

3. in environments where the traffic per site is low compared to the transport capacity, add-drop multiplexers (ADMs) provide a cost-effective port aggregation solution.

However, the extension of the ring topology to the network itself has usually been restricted to traditional TDM transports:

1. TDM transports have their genesis in the days of narrow-band services and expensive optical transmission, and so the ADM was a cost-effective method of combining aggregation and transport functions;

2. the rigid containerization of TDM transport meant that for nearly all such technologies, protection bandwidth had to be preassigned, and so this bandwidth was "stranded" during normal operation. On a ring topology,

(Continued)

this permitted simple, and very rapid, data plane protection mechanisms. The 50 ms fail-over time of SONET/SDH became a "gold standard"; its absolute necessity may be questioned, but it could be done.

The packet environment has evolved with a very different set of axioms:

1. the uptake of broadband services often results in a single access point requiring a dedicated fiber or wavelength for backhaul, and so the ADM ceased to have the same network value. The duct topology may remain rings, but the installed optical capacity inside it is p2p;
2. the flexibility of packet forwarding allows flows to be multiplexed in a reactive manner, with no need for hard-bounded preallocation of resources. As a consequence, the packet networking model endeavors to make use of all available bandwidth at all times, accepting more or less degradation under fault as a function of overprovisioning, an operator-determined network parameter;
3. Shortest path first routing technology allowed the implementation of distributed control planes, which enabled the application of flexible packet forwarding to an arbitrary mesh topology, and allowed restoration from faults to make use of whatever network capacity remained.

Although there has been recent work on ring protocols, it appears that this is becoming a "minority sport." The fundamental reason is that the relative simplicity of ring protection is only achieved at the expense of a major restriction; no extension of connectivity beyond a ring can be accommodated (e.g., dual-homed access to a ring is precluded). By contrast, a ring is simply a special case of a mesh, and can be handled without exceptions within a general mesh environment. When the ring was a widespread special case, it can be held that it warranted optimization; when it ceases to be so, it is not.

The Provider Core Network

As indicated above, provider core networks largely carry highly aggregated IP/MPLS services, and these are forwarded based upon IP routed paths with selective use of IP/MPLS TE. Since the routers exist and the traffic they emit is highly aggregated, there appears to be little merit in a layer 2 underlay for such networks.

The area where there is likely merit is "IP/MPLS router bypass." In a carrier environment, a significant number of enterprise services are long distance p2p, now increasingly with Ethernet presentation to the customer. This is particularly true in the regulated environments mentioned above, where backhaul of residential broadband access to a significant number of ISPs can approximate a national-scale networking issue.

Such services were traditionally carried directly over SONET/ SDH, with no necessity to carry them over a routed platform, and this decision both controls the number of router ports and straightforwardly allows a "gold standard" Service-level agreement (SLA) to be offered.

With PBB-TE inheriting the SONET/SDH attributes and applying them to packets, the same p2p service capability can be offered by this technology too.

The Data Center

Until now, data centers have been built using enterprise network practice, with both Ethernet and IP routing. With nonstop operation and dual homing of edge switches a key requirement, the extent of Ethernet bridging has typically been limited to no more than two levels of aggregation switching before traffic has been terminated on a router. This produced a unique market for huge switching platforms, so that the disruptive consequences of the Spanning Tree Protocol (STP) under failures can be limited or even eliminated.

For most data centers dedicated to a single enterprise, this model at the time of writing remains adequate. However, with the rising awareness of global warming, there has been increasing focus on data center virtualization and the concepts surrounding cloud computing. This is true for both enterprises such as Google which offer Web-based services on a global scale, and service providers offering multiuser hosted IT facilities. A server blade dedicated to a single application is widely reported to be only 15–20% loaded on average, even during the business day (see, e.g., [Bathwick]), and so there is a substantial energy win to be had from offering numbers of virtual servers per blade, individually assignable to applications, and supported by the ability to dynamically manage and distribute load across the server community.

This requires not only server configuration, but also the configuration of IP subnets to match the (now) distributed virtual servers and the IP addresses assigned to them. In this application, the SPBM technology outlined earlier offers a very good match:

- a set of service identifiers (I-component Service Instance Identifier [I-SIDs]) defines a perfect virtualized emulation of an Ethernet LAN segment at layer 2, which **is** an IP subnet at layer 3;
- I-SIDs need configuration at the edge (endpoint) only, after which the routing system installs the rest of the state; this is a very robust scheme well suited to automation in conjunction with virtual server configuration;
- SPBM eliminates the spanning tree protocol and inherits a link state property key for a high availability environment, which is that only the traffic actually using a specific resource is affected when it fails;
- SPBM offers far superior utilization of the connectivity patterns of emerging data center architectures such as "fat trees" than was achievable with spanning trees.

The scaling of very large data centers is providing a whole new set of challenges which the industry is struggling to address. Server virtualization means that a single rack can host many hundreds, sometimes thousands, of IP terminations. Regrooming of virtual server instances within a flat layer 2 network means that the relationship of identity (MAC) and location (IP) has become inverted, with the IP address becoming the identity of a virtual server instance, which has mobility within the data center network, and the MAC defining the location of the physical server blade. Finally, the switching architectures to scale the bandwidth within the data center, such as Clos or "fat trees," use highly parallel connectivity, and so are pushing the boundaries of multipath forwarding.

Traffic Patterns and Service Models

From the sections above, discussing the characteristics of different networks, it will have become apparent that the traditional enterprise environment is in general distinctively different from carrier networking:

- there is a natural hierarchy, from desktop to core;
- except within very large modern data centers, fine-grained virtualization is not often seen, and network partitioning is typically at the level of workgroups of significant size;

- it is usually cheaper to "throw bandwidth at the problem," rather than concentrate on maximizing the **efficiency** of bandwidth use.

As a result, the next few sections are deliberately more focused on carrier network concerns; this appropriately mirrors the major motivation behind SPB, which was to **extend** the range of applications accessible to Ethernet technology, not merely to enhance its capabilities in its traditional strongholds.

A major challenge in networking is maintaining a design balancing act, because there is a crossover point between the simple "aggregate, backhaul, and switch" of a hierarchical network, and the use of multipoint and partially meshed connectivity which actually minimizes the overall required bandwidth of the network. This is because:

- "aggregate and backhaul" eliminates **spatial** variations in offered load by individual users (often called "slosh," at least by the authors, as it is analogous to rocking a shallow tray of water). This is because traffic is being passed up a hierarchy on a single known path, switched, and returned down another known path, but this predictability is at the expense of some traffic being sent further than strictly needed;
- However, when the traffic on a mesh network has been sufficiently aggregated such that the cumulative spatial variations in offered load by users are averaged out, the degree of dilation (the ratio of trunk capacity to the average load carried) required for a given level of service is minimized.

The use of "aggregate and backhaul" has a further advantage that it constrains the size of the domain where network behavior is entirely statistical, in addition to driving up the size of "N" (always a good thing in statistics) to reduce the impact of traffic burstiness on the links. The behavior of the backhaul portion of the network is very predictable, while the meshed core network makes efficient use of the available capacity, with the magnitude of the statistical effects being mitigated by a high degree of aggregation.

A separate variable in this equation is the legal intercept requirement which many carriers face. This typically requires that all traffic be brought back to a point where wiretapping capabilities are deployed, this functionality being colocated with service nodes.

SIDEBAR: *Virtualization Models and Security*

When considering virtualization, the partitioning of infrastructure to create closed communities of interest, we can consider multiple models:

- a peer model, where the community of interest peers with the network, and distributed forwarding and filtering policy dictates the constraints of connectivity applied to the community of interest, or
- an overlay model, in which the application of policy primarily occurs at the edge of the network.

The challenge for most carriers is to provide large-scale virtualization, because the number of access points in an individual customer virtual network is typically a fraction of the total number of points of presence of the carrier network. Hence, most successful carrier architectures seek to minimize the number of points in the network that need to be "touched" to perform adds, moves, or changes to any individual virtual network.

When looking at the gamut of technologies deployed for Ethernet services, we see a mixture of peer and overlay models. Overlay is considered to be the more scalable approach; however, multicast in particular gains no scalability in an overlay environment. This is because in a network of any size, the number of possible sparse multicast trees rapidly becomes huge, and the probability of requiring multiple instances of any specific tree, which could be aggregated and so benefit from hierarchy, becomes vanishingly small.

Ethernet Q-in-Q is a peer model where the application of the S-tag identified the community, and configuration of S-VLAN connectivity across the core constrained the set of access points with which a community member could exchange data. It does have the virtue that an S-VLAN tag is the unit of "isolation," and hence the actual provisioning of service isolation required is aggregated and therefore comparatively trivial. However, SPBV has repurposed the "per-service" interpretation of the S-Tag, and uses the S-Tag instead as a per-source tree identifier. Consequently, SPBV lacks even the limited virtualization support of Q-in-Q bridging.

MPLS VPLS is an overlay model, creating a split horizon mesh of p2p pseudo wires for each VPN between the individual VPN access points. During modifications to a given VPN instance, no core infrastructure is touched, but any modification must be reflected in the state at every access point of that VPN.

SPBM scaling ends up as a hybrid of peer and overlay, due to the "flat" control plane required to implement the per-service multicast capability. The unicast component is aggregated and uses an overlay model, but the multicast component cannot use this model, for the reason summarized above.

Network Efficiency

P2p services are straightforwardly engineered, especially when aggregated and backhauled, but the implementation of multipoint services is a much more complex matter. The bursty nature of data traffic can be exploited by overbooking shared capacity, and relying on statistical variations in individual customer demands to ensure that actual congestion is rare. For such contended p2p packet services, the primary influence of the customer on the network is "how much" bandwidth, constrained to a predefined path, whereas for multipoint the additional and significant dimension of "to where" is added, with the potential impact of uncontrollable customer behavior being greater the larger the size of the set of "to where" possibilities.

Multipoint connection implementations are sufficiently complex and stateful that operating on aggregates (with many different connections sharing a common routing solution) has been required to achieve the combination of both scale and resiliency required to generate practical service offerings. Resiliency at scale is a particular challenge, because the response to a fault in one segment of a multipoint connection depends on the whole topology and the exact location of the fault; there is no "one size fits all" response analogous to the protection switch widely deployed to spare p2p connections. The use of a routing system to control a destination-based forwarding plane is the most scalable form brought forward so far in the industry, and this is the model used by SPB.

However, critical examination demonstrated that the distribution of shortest path routed traffic was relatively fragile and unstable when it was perturbed, either by a real failure or by an "artificial" topology change such as a link metric adjustment. This is certainly true if anything more than "best effort" connectivity is the goal. Attempts to modify the distribution of the traffic matrix via metric manipulation can have large and unpredictable results. This is especially noticeable when statistical load spreading in the form of equal cost multipath (ECMP) is employed in arbitrarily structured networks. The conclusion which emerges from examining network behavior is that whatever technique is used, ideally, a mechanism is needed that does not modify the distance between any two points in the network. In this way, the unpredictable consequences of metric modification can be avoided. However, this does require path generation techniques that can take advantage of breadth of connectivity between any two points.

An example approach to efficiency optimization might be to observe that if each hop identified all the paths of equal distance between itself and a given destination, and had knowledge of the capacity of each path, and could perform a weighted assignment of load, then the network could be infinitely and predictably tuned via modification to the advertised capacity. However, the actual quantity of state required in the forwarding engines to implement such a scheme (a weighted forwarding table per destination) suggests a practical implementation is intractable.

SIDEBAR: *Equal Cost Paths and Trees: Example*

It is worth clarifying the load-spreading technique currently used by SPBM, because it is distinctly different from the widely deployed ECMP, and has different attributes.

ECMP is illustrated in Figure 2.2. Each node identifies all of the equal cost next hops from it to a destination, and spreads the traffic over the set of its egress ports corresponding to the next hops. The spreading operation typically uses some form of hash function, which should be unique to each node, applied to the packet address fields modulo the number of next hops to give a next hop selector. Uniqueness is required to avoid the traffic arriving at a subsequent ECMP hop being precorrelated, when the quality of randomization suffers dramatically. This is a local (hop-by-hop) function, and as can be seen from the figure, paths can reconverge at multiple ingress ports of intermediate nodes, and the aggregate traffic can then be respread over multiple egress ports.

The attraction of this technique is that it statistically spreads traffic over all of the eligible network resources, with only a modest state burden, and minimal

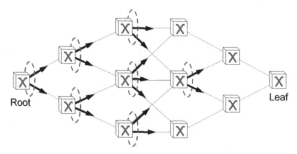

Figure 2.2   Hop-by-hop load spreading—ECMP.

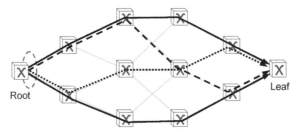

Figure 2.3   End-to-end load spreading—ECMT.

management intervention. Consequently, it is ubiquitously deployed in the Public Internet, to which its properties are well matched.

ECMP does present issues. It cannot be applied to multicast, so congruent forwarding of unicast and multicast frames is a property that cannot be preserved if ECMP is employed. It assigns traffic purely on the basis of a local view, inferring equal capacity to each of the set of available next hops, something that is true of highly regular networks such as those found in data centers but not true in arbitrary deployments. It also is the antithesis of proactive operations, administration, and maintenance (OAM) because of its statistical properties. When the payload content is the source of entropy for hashing, it makes it impossible to identify a specific service or customer flow with a particular route through the network for instrumentation purposes, without impersonating it. This makes OAM challenging as the OAM design needs to somehow proxy for the payload. More fundamental from the perspective of SPBM, the same attribute means that the go-return path congruency property is lost: This property is fundamental for SPBM because loop mitigation in the form of reverse path forwarding check (RPFC) relies on a single path from any point back to the root of a tree.

As discussed in the main body, the ability to utilize breadth of connectivity is important to use capacity efficiently in a routed network. SPB has the challenge of utilizing network breadth while maintaining symmetric congruency between unicast and multicast. SPB therefore uses an end-to-end load-spreading technique, known as equal cost multiple tree (ECMT), which is illustrated above. In this, end-to-end paths are computed, each by a different tie-breaking variant of the routing algorithm, and associated with different VID planes. A subset of the possible paths is shown above. The root assigns traffic to a VID plane, thereafter the forwarding is completely deterministic, and has a true p2p model within any VID plane, thus delivering the required congruency property. Another attractive attribute of this is that the Ethernet OAM tools become usable "as is." A further desirable property is that in failure scenarios, unlike ECMP, the distribution of traffic remote from any failure is unperturbed.

Traffic Engineering and SPBM

When traffic engineering (TE) is considered, two modes of operation come to mind. The first is entirely proactive, often termed "design and assign," in which the exact placement of the traffic matrix for a given service is planned in advance in the light of the existing network loading, and then explicitly and individually instantiated. The second mode is to map customer traffic onto existing aggregates and retroactively fix hot spots as they occur, what we term an "observe and react" mode of operation.

Network scaling suggests that operating the first mode of operation "entirely flat" has clear limits as absolutely all of the required state is incrementally added to the network, as an exception to any sort of "aggregated" solution. The amount of state in the core of the network is in proportion to the number of services.

This highlights that what a routing system does best is to work with aggregates, hence bounding the amount of state in the network on that basis. The routing system is considered to be the primary system for the distribution of load in the network, the mode of operation is "observe and react," and the routing system itself is used make adjustments when the current routed solution does not match the offered load, which is usually identified by the utilization of a given link exceeding a preset threshold. The solution is to provide some form of "exception" that moves traffic away from a hot spot and onto less utilized parts of the network, but otherwise causes minimal perturbation on the parts of the network that "ain't broke."

Experience suggests that the only useful exceptions for TE purposes are those that add virtual links (in some form) within the network, where these virtual links are still considered to be topological components by the routing system. This is because for connectionless routing, it is only by artificially increasing the mesh density can any degree of subtlety in manipulation of a routed traffic matrix be achieved and other significant properties conserved. What this does mean is more state, less optimal forwarding of traffic (except for the desired "sweating of underutilized assets"), and slightly increased convergence times as the number of links has been artificially increased. When this mode of operation is used, the decommissioning of exceptions as new capacity is added to accommodate network growth is beneficial in terms of both minimizing the overall volume of network state, and reducing convergence times. The good news is that networks are not getting smaller, and so while an initial build of a network may not correspond to offered load, the situation will self-correct over time.

These considerations set the basic direction for the extensions to SPB discussed later in "Extended Connectivity Models: Nonplanar Graphs" (p. 181).

Multipoint Resiliency

In its fundamental principles, resiliency for basic **p2p** connections is straightforward, and is a trade-off between responsiveness and preinstalled state. This is a consequence of the decision as to whether the responsibility for executive action to recover from a failure has been delegated to the data plane or the control plane:

- restoration can, as in conventional IP, be left to the control plane
- protection paths, either at the section or end-to-end level, can be preinstalled. This is tractable because any fault on the "working" element results in the same action; "use protection path" (singular). This is a rapid low-level action, and scales because only the directly affected network elements are involved.

Fast resiliency for **multipoint** connectivity is as desirable as for p2p connections, and yet it has been an elusive goal. The basic reason is that, for an arbitrary tree, the network response to any particular fault will be unique. The portion of the tree to be recovered is entirely dependent on the set of leaves impacted, in contrast to the single common consequent action to the set of all possible failures in a given p2p path. The closer to the root that the failure occurs, the larger the set of leaves for which connectivity must be restored. Furthermore, for other than a spanning tree or rooted multipoint tree, the number of trees to implement multipoint connectivity, and which therefore require simultaneous restoration, can be quite large.

Consequently, restoration of multipoint connectivity has relied on a control plane for anything other than completely trivial and constrained topologies. Ethernet traditionally has used the STP, which after detection of a topology change blocks all ports to user traffic until a new spanning tree has been robustly negotiated. In the IP world, although link state routing converges the unicast topology view substantially faster than STP, multipoint restoration is still slow, because reinstallation of multicast state, by for example Protocol Independent Multicast (PIM), requires hop-by-hop signaling running over the unicast topology **after** it has converged.

SPB breaks new ground in using only the output of computation over a link state topology database for the generation of both unicast and multicast forwarding tables, obviating the need for multicast signaling. It further allows a "multipoint service failover" model, described later under "Provisioned Trees with Routed Backup" (p. 181).

HISTORY

This document does not attempt detailed coverage of the fundamentals of Ethernet bridging, nor does it have anything to say about the Ethernet physical layer; and readers are referred elsewhere for this, for example, [Spurgeon]. A detailed treatment of networking protocols at both Layer 2 and Layer 3 can be found in [Perlman]. For this document, history means "recent history," starting with Provider Backbone Bridges (IEEE 802.1ah). [PBB] radically increased the size of the identifier spaces associated with bridging, while keeping a common technology base and a spanning tree mode of operation. The latter represents a mismatch with the desired applications and scale, which SPB's application of link state routing addresses.

PBB is a key antecedent of SPBM, which introduced basic concepts that SPBM uses, and which were eventually also harmonized with SPBV where appropriate:

- PBB introduced true hierarchy to Ethernet for the first time, with customer Ethernet traffic using C-MAC addresses being encapsulated in Backbone (B-MAC) addresses across the backbone. This has a number of benefits; as well as state reduction in the core PBB domain, **all addresses in the backbone are known to and under control of the network operator.** This is because C-MACs are encapsulated, and therefore hidden, and B-MACs are all associated with the network operator's switches in the

PBB network itself. In other words, this backbone address information can be known as part of network operation a priori, and if B-MAC reachability information can be distributed somehow, the need for MAC learning in the core (which PBB uses) can be eliminated (which is what SPBM does).

- PBB introduced a quite distinct service identifier field, which is cleanly separated from the backbone MAC addresses and backbone VIDs, and this uniquely identifies a network-wide service instance. This distinguishes PBB from its predecessor, Provider Bridges (PB, or IEEE 802.1ad), in which an outer S-VLAN tag was available to providers to allow their VID space to be administered separately from their customers' VID spaces. This nonetheless leaves providers with an overloaded field, with the S-VLAN tag identifying not only a customer, but also a virtual network topology instance. This overloading and the modest size of the VID field results in substantial constraints.

- PBB preserves the per-service scoped multicast attribute of PB, by substituting the Provider VLANs (S-tags) with backbone multicast MAC addresses, which achieves the same goal without the PB scaling constraints.

The key new functionality to achieve this is known in PBB as the I-component. From [PBB], "each I-component is responsible for encapsulating frames received from customers and assigning each frame to a backbone service instance. The backbone service instance consists of a set of BEBs [Backbone Edge Bridges] that support a given customer's S-VLANs, and it is uniquely identified within the Provider Backbone Bridged Network (PBBN) by a backbone I-component Service Instance Identifier (I-SID). The customer frame is encapsulated by an I-TAG, which includes the I-SID, and a set of source and destination backbone MAC addresses. The backbone MAC addresses identify the BEBs of the backbone service instance where the customer frame will enter and exit the PBBN. If the I-component does not know which of the other BEBs provides connectivity to a given customer address, it uses a default encapsulating backbone MAC address that reaches all the other BEBs in the backbone service instance. Each I-component learns the association between customer source addresses received (encapsulated) from the backbone and the backbone source address, so subsequent frames to that address can be transmitted to the correct BEB."

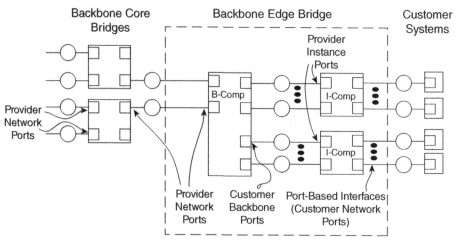

Figure 2.4   Port-based interface structures (from [PBB]).

The figure above taken from IEEE 802.1ah [PBB] illustrates one option for the inclusion of the I-component into the Ethernet architecture and shows the multiple layers of demultiplexing that can happen within a BEB. The first level of demultiplexing is achieved by using B-MAC addressed provider instance ports, and the next level of demultiplexing, to the individual service, occurs on the basis of the I-SID within the B-MAC addressed I-component instance. In the example above, the service is associated with the port; for another interface type, it is associated with an S-Tagged VLAN on the port, to interface directly with Provider Bridges.

PBB-TE [PBB-TE], originally known to the industry as PBT, exploited the complete knowledge of backbone addressing within the backbone domain, which is a key consequence of the PBB hierarchy, combined with the emerging OAM standards 802.1ag and Y.1731. With an alternative way of learning backbone network topology and making it available to a management system (using Connectivity Discovery, IEEE 802.1AB), there ceased to be any need for conventional "flood and learn" of B-MAC addresses. Instead, "flood and learn" is turned off in PBB-TE, and forwarding tables are explicitly configured by the management system to build routes which have been planned within the management system. Further enhanced robustness follows from the complete knowledge of all routes by the management system, because it is thereby known for each port on a bridge what destination B-MAC addresses are expected and are valid. So, receiving an unknown B-MAC is indicative of a configuration error, whereupon the frame is discarded instead of flooded, and its arrival can be alarmed as an error condition.

The elimination of flooding from the Ethernet equation and the shift to explicit population of the filtering database (FDB), either by management or a control plane, had very important implications for the evolution of Ethernet. The key implication is that a simply connected and heavily constrained active topology (a spanning tree) was not required for the PBB-TE mode of operation. This is because loop avoidance in the data plane was not required, because it had moved to being a control or management plane problem. In essence, all ports could be unblocked, and for the first time all links in the network could be exploited for the carriage of traffic.

When the dust settled, what PBB-TE revealed was that Ethernet had a massively scalable data plane design (it is possible in theory to fully connect some $2**46$ end systems for both unicast and multicast 4094 times before exhausting the data plane identifier space), and that lurking within most modern Ethernet bridges was a massive MAC/VID cross connect, the control of which could be simply exposed and exploited. The other key part of the toolkit was that, as a consequence of an independent industry standardization initiative, Ethernet had finally attained a comprehensive suite of OAM tools.

Even with these powerful antecedents, SPB has been through various incarnations since its inception. It is the result of several years of refinement, periodic parking, and subsequent revisiting; the latter has often been triggered by unexpected requests of the form "could it do X for us?" from prospective users.

SPB started simply with the concept of applying the PBT technique of configuring bridge forwarding tables but with the modification of this being driven by a link state routing system. The original idea was to leverage Ethernet properties to produce an MPLS-LDP replacement, as part of the thought processes associated with exploring whether Ethernet could be equipped to compete as a lightweight MPLS.

An overarching requirement was to replicate the "minimal intervention" paradigm of PBT. PBT required only a small change to the existing standards, the enforcement of configured behavior by discarding unknown B-MAC addresses, and therefore had high value in proportion to the standardization effort, and at the same time low barriers to industry adoption. Our design intent for any follow-on enhancements to Ethernet was similarly to require only minimal changes to the standards, and with any necessary changes having high value, such that the time to implement them in commodity silicon for use in the data plane would be minimized.

The earliest starting point was noting the similarities of our requirement to the layer 2 address resolution embodied in Juha Heinanen's proposal for ATM/IP integration [Heinanen], whereby routers advertised ATM end system addresses (AESAs) using the routing system and employing an opaque LSP. In essence, address resolution became a routing system function. This meant that peers could then set up switched virtual circuits knowing the domain-wide layer 2 identifier of the destination router. This obviated the need for complex and failure-prone address resolution overlay architectures such as classical IP over ATM (CLIP) and Next Hop Resolution Protocol (NHRP). As in the [Heinanen] proposal, there was also a routing protocol involved in our model, and so we too could avoid the problem of N^2 routing adjacencies.

It was realized that MAC addresses associated with IP addresses could be simply flooded in the routing system, with no need for hop-by-hop personalization (because they are network-wide), and that transit nodes simply needed to populate their forwarding tables directly with the MAC addresses learned from the routing system. The fact that destination layer 2 data plane information of domain-wide significance could be carried in the routing system meant that both layer 3 to layer 2 address resolution, *and* the requisite layer 2 signaling to set up paths could be condensed to a single control protocol without resulting in N^2 routing adjacencies. When the routing system converged, it simply placed the advertised MACs into the forwarding table and connectivity "just happened." In essence, any device on the edge of such a cloud had its Address Resolution Protocol (ARP) cache prepopulated, which is exactly how one prototype was implemented.

At the time it was expected that loop mitigation would be addressed by other stardardization efforts that were then underway, one candidate being the Internet Engineering Task Force (IETF) effort known as "Transparent Interconnection of Lots of Links" (TRILL). It should be noted that TRILL was focused on Ethernet replacement, while we at the time were focused on an MPLS replacement, and the subsequent convergence of the target client application, enterprise Ethernet, was a small irony. However, the solution emerging from TRILL at the time was seen to be a bastardized MPLS label, and it retained the IP time-to-live (TTL) field for loop mitigation. TTL for loop suppression in the Ethernet environment did not preserve aspects of the Ethernet service model, as it would allow for frame duplication, and we predicted that in the fullness of time it would require duplication of all the associated tricks and tools that experience with IP had ultimately deemed necessary. So consistent with the design principle of minimum intervention, another solution to loop mitigation was sought.

The solution adopted was to adapt MAC learning to perform the role of MAC policing as a loop mitigation mechanism. In normal MAC learning, if the source address of an ingress frame is not present in the filtering table, an entry is created pointing to the ingress port. In SPBM, the filtering tables are populated by the control plane; consequently, the arrival of an ingress frame without a filtering table entry can be considered to be an error condition, and the frame should be discarded. This was a solution that required minimal changes to bridge implementations, and no "on the wire" modifications (which would be required to add, e.g., a time-to-live field); hence, it would be a desirable solution from the standpoint of both adoption and standardization.

There were a number of serendipitous consequences of this design decision, the full implications of which were only later fully appreciated. First, what was needed to achieve the requisite connectivity was a **connection-oriented** routing system; the key distinction between connection-oriented and connectionless is that "come from" as well as "go to" needed to be known and policed:

- although the forwarding process retained Ethernet's destination-based model (and so also its very desirable scaling properties), the ingress check also ensured that frames were on the path from the source to each intermediate node on the path;
- this combination of properties essentially defines a connection.

This is discussed in detail in "The CO Nature of Ethernet and its Implications for Routing," see p. 126.

Notice, however, that the use of two shortest paths between an intermediate node and a pair of endpoints says nothing about the shortest path between the endpoints themselves; the sum of the two shorter sides of a triangle is in general greater than the longest side. This is obvious when considering p2p connections. The possible pitfalls become more insidious when considering multicast, see "Recasting the Group Multicast MAC Address for Shortest Path Trees" (p. 66).

The RPFC technique, although adding significant robustness to loop mitigation, was not perfect, and methodologies were discovered that permitted looping scenarios (however implausible) to be generated at will. From this emerged a combination of control plane and data plane techniques that had the necessary "belt and suspenders" attributes.

The required authoritative scheme, based on topology database synchronization as part of reconvergence, is described in the Control Plane chapter under "Loop Avoidance in SPB" (p. 111).

Recalling that the original thinking had focused on unicast for layer 2/layer 3 integration, it was clear that RPFC actually had some downsides in that application:

1. It required symmetrical metrics and a single shortest path between any two points, which is hardly state of the art for connectionless routing.
2. It was overly aggressive in squelching traffic while the control plane was reconverging after a topology change.

One approach considered was that RPFC would be turned off during periods of network instability so that nonoptimal forwarding would be permitted, and turned back on after some period of time to squelch any persistent loops. Later insights, also described in the "Loop Avoidance in SPB" (p. 111) section of the Control Plane chapter, showed that looping was more likely during control plane reconvergence, and that RPFC should not be turned off, at least for multicast which absolutely required "break before make" behavior for robustness. For multicast trees, the authoritative loop prevention technique alluded to above was also developed, with RPFC as an "always on" safety net to counter implementation errors or persistent hardware faults.

However, a benefit of using RPFC as one mechanism for policing loop freeness was that a common VID would be required for both directions, so that an IVL (Independent VLAN learning) bridge implementation, already long standardized, was easily adapted to RPFC, and if we pursued load spreading using multiple topologies (one per VID), then RPFC would still be able to resolve correctly which topology it was policing.

The final component came from the ongoing PBT work and the trivial and obvious realization that multicast MAC forwarding could be configured in the same way that unicast MAC forwarding was configured. At the same time the IEEE 802.1aq shortest path bridging work was focused on symmetric shortest path forwarding and edge rooted spanning trees for the Q-in-Q forwarding plane, with each edge rooted tree identified by a VID (SPBV). This suggested that a small extension to the concept of link state-driven layer 2/layer 3 integration would simultaneously permit the creation of congruent mp2p unicast and p2mp multicast trees to and from any given BEB, with the trees now identified by B-MAC addresses. At this point the role of SPB was being recast, from a unicast layer 2/layer 3 integration mechanism to what it is today, a means of virtualizing large numbers of LAN segments in a "better" PBBN.

At this point, the clarity of hindsight showed us where we would actually end up; what was first introduced to the industry as Provider Link State Bridging, and then became SPBM, actually rounded out the overall Provider Ethernet toolkit. Very simply, there are two connectivity bookends:

- there is the traffic engineered p2p virtual connection, which is what PBB-TE delivers;
- there is the multipoint LAN segment, which is what PLSB/ SPBM virtualizes.

Extending those concepts into layer 3, we find the unnumbered link and the virtual subnet, described later in "Introduction to IP/SPBM Integration" (p. 164).

Similarly there are two models of operation:

- comparatively fine-grained placement of traffic paths to the point where a distributed control plane would be overwhelmed by the associated state, and therefore delegating responsibility for resilience to the data plane was pretty much mandatory, again embodied in PBB-TE;

• the manipulation of aggregated multipoint connectivity relying on the control plane for resilience, embodied in SPB.

In the final step, the original design intent in many ways went nearly the full circle. Later, under the name "Layer 3 Integration with SPBM" (p. 163), also known as "IP Shortcuts," we will describe a common layer 2/layer 3 control plane for the forwarding of IP over Ethernet, which is exactly the function which IP/MPLS performs. What had changed since the start of this voyage is that, in the case of SPBM, there was now a whole raft of virtualized layer 2 capability to be exploited which had not been envisaged at the outset.

LYNCHPINS: CONSTRAINTS WE CHOSE TO RESPECT

Here, we briefly discuss two properties which are essential underpinnings of conventional learning bridges. In moving from a "flat" spanning tree environment to a client–server model using routed shortest path trees in the server layer, it would have been possible to lose these properties from the server layer. Instead, we made a conscious decision to preserve them, despite the challenges they presented in routing. Having made these decisions, a third general "minimum intervention" principle follows.

The Benefits of Symmetry and Congruency

Once our application focus for the SPBM technique changed from layer 2/layer 3 integration to being predominantly L2VPN in the form of virtualized LAN segments, the requirement for symmetric metrics and congruency in order to use RPFC switched from a limitation to an asset, because of the much more stringent loop avoidance requirements of the layer 2 multicast environments.

Conventional Ethernet relies on symmetric forwarding and unicast/multicast congruency for a number of reasons. In summary, bridging currently assumes and depends upon symmetrical physical links between bridges, to ensure that MAC learning works, that spanning tree converges, and for other reasons, and if these properties are preserved when virtualizing bridging good things follow. The principal properties are:

1. No frame misordering in race conditions (during flooding and learning scenarios);

2. Symmetrical fate sharing avoids the disruption of customer (C-MAC layer) STP exchanges, caused by introduction of uni-directional modes of failure which fall outside the design requirements of spanning tree;

3. Congruency is maintained between forwarding and customer layer OAM flows, which is important because customers will likely use 802.1ag multicast addressing for end-to-end OAM, and will wish to be assured that this is probing the unicast path as well.

So, ensuring the PBN and PBBN used symmetrical paths and unicast/multicast congruence now became a virtue from the point of view of the supported service model, and serendipitously permitted RPFC to become a loop mitigation mechanism.

Benefits of Self-Describing Frames

A key design property of Ethernet frames is that they are fully self-describing. Not only does this enable a much simplified control structure for Ethernet, it has OAM advantages as well. No signaling system is needed to bind a link-local identifier (otherwise called a label) to the global endpoint address because this address travels in the frame. When compared to label swapping, there is one less level of indirection in the forwarding plane, which means the number of modes of failure has also been diminished and the amount of state to be synchronized has been correspondingly diminished.

Minimum Intervention

As in the case of PBT, SPB was about minimal intervention in the Ethernet data plane. PBT had obtained traction precisely because it was seen as the addition of new properties to an **existing** technology, **not** the invention of a new one, and we wanted to maintain this completely justified perception. The explicit change needed for SPB was the addition of asymmetric VID translation for SPBV, the addition of RPFC to the existing PBT profile of behavior for SPBM, and the ability to use a per I-component multicast address where the OUI identified the tree root instead of being a well-known reserved value. This (S,G) address actually allows existing filtering to perform the function of RPFC for multicast DAs, but the SA must be inspected for RPFC on unicast.

This approach also ensured that other important attributes such as split horizon forwarding were preserved, meaning that a frame will not be forwarded back onto the interface of arrival. This is a desirable property as the "one hop micro-loop" scenario could be removed from consideration when examining loop avoidance. It cannot be emphasized enough that the Ethernet community is highly worried about looping, because of its "meltdown" consequences with multicast traffic, as a result of the small physical scale and consequently very low latency of typical Ethernet deployments. The STP is the Ethernet "gold standard" because, despite its operational disadvantages, it does guarantee a loop-free environment, and possible alternatives are judged by comparison.

REINTERPRETING WHAT ALREADY EXISTS IN ETHERNET

The key word here is "**reinterpreting**"; we are **not** redefining any Ethernet concept, merely revisiting their implicit connotations, but always within existing definitions. For example, "VLAN" conventionally carries the implication of a subsetting mechanism, intended to limit the scope of a broadcast domain within a network. SPBM is able to exploit the separation of VLAN from a strict subsetting function implied by PBB, such that a VLAN becomes a construct which defines a single set of trees covering the entire network, with multiple VLANs being assigned and used in a coordinated fashion to provide multiple meshes over the same network. Both are entirely valid interpretations of the IEEE 802.1Q definition of a VLAN, but the latter has not been widely employed before.

So, we now consider the VLAN, unicast MAC addresses, and multicast group addresses, and elucidate the interpretation which SPB has put upon them.

The Meaning of the VLAN in PBB-TE and SPBM

Both PBB-TE and SPBM subtly recast the role of the VLAN in Ethernet. The VLAN originally resolved to a set of ports in a spanning tree, then with the addition of multiple spanning trees it resolved to identifying a set of ports in a spanning tree instance. Finally, with both PBT and SPBM unblocking all ports for the associated VID within a provider backbone network, the VLAN has now become a backbone data plane instance identifier. An alternative way of expressing this is to view the VID as identifying a single forwarding plane which has the capability of fully connecting all SPBM nodes in the network. Diverse routes between endpoints are supported by multiple equal cost trees (ECT), each created using different tie-breaking algorithms in IS-IS, and are identified by different VIDs in the forwarding plane.

Although one can envision multiple paths between any two MAC endpoints within a single VID, the constraint required to ensure that a given FDB only had single egress port for any B-MAC presented would be that the paths were guaranteed to be completely diverse end to end. This is an artificial constraint which could not be sustained across failure scenarios and one that had no value when one could simply use a different VID per path permutation, and so remove any diversity constraint from path planning. Further, to produce multiple paths and full connectivity respecting the same constraints would be pretty much impossible. Similarly we would not assign multiple MACs to a BEB simply to permit multiple paths in a single VID; we keep the clean architectural separation of the endpoint identifier (the MAC address—"where") and the path instance identifier between endpoints (the VID—"how").

Duality of the VLAN: A Planar Network Instance or a Virtual Link

The VLAN is an interesting networking construct. The term LAN is a historical artefact of the notion that it was a broadcast domain associated with a spanning tree instance. We can now reconsider it as identifying an instance of a virtualized network with its forwarding plane, where the definition and utility have been broadened. However, the subsetting capability embodied in the original VLAN implementation still exists, and the logical bookends of this concept are now that it

- can refer to a complete networking instance, as used by SPBM, or
- be constrained to a virtual p2p path constrained to a single link (see e.g., "Topology Modification for Traffic Engineering Purposes", page 183),

and where the rather large space in between is currently uncharted territory.

As a networking instance identifier, one can also infer the forwarding and operational behavior associated with that instance, as SPBM, or PBB-TE, or conventional bridging, to allow "ships in the night" operation of different behaviors on the same infrastructure. We can also translate VIDs symmetrically at port boundaries, which suggests that the forwarding of a frame across a network could transit a composite hybrid of modes of operation and filtering styles if desired.

Prior to SPB in existing Ethernet standards, only bidirectional VID translation was permitted, for the purpose of decoupling VID assignment on interfaces at domain boundaries from the assignment used within a domain. Asymmetric translation is required for SPBV and this will facilitate certain aspects of the evolution of SPB, notably multiarea topologies and TE, which can be more simply implemented if asymmetrical translations of VIDs are allowed. This is primarily so that a single forwarding fabric can integrate multiple forwarding domains without artificial partitioning. These scenarios are identified later in the document ("The Unary FDB Problem at SPBM ABBs," p. 176).

MEANING OF MAC ADDRESSES IN PBB-TE AND SPBM; PORT AND NODAL MACS

PBB is able to support more than one B-MAC associated with the internals of a node, depending upon the implementation model adopted. So the addressing can be at the granularity of node, line card, network processor on line card, UNI port, and so on. In the language of the PBB Standard, from which Figure 2.4 was reproduced earlier (in "History," p. 52), an individual backbone MAC address is assigned to each Provider Instance Port (PIP).

When an I-SID is associated with only a single element on a node, the I-SID itself is sufficient information for the NNI to demultiplex the service to the appropriate Customer Network Port and thence I-component. Where the implementation of an I-SID is spread over multiple components in the node, it may be chosen to provide a unique B-MAC address for each component. This choice represents an implementation trade-off; a unique address for each component clearly impacts scaling, but allows the NNI to switch on the B-MAC only, whereas if a nodal B-MAC is used, the NNI has to implement a Virtual Switch Instance (VSI) at the C-MAC layer for every service with more than one I-component.

This means that in IS-IS for SPBM port MACs are advertised:

1. with an explicitly associated B-VID, so that the port MAC does not need to appear in every B-VID in every FDB;

2. with explicitly associated I-SIDs which are to be forwarded using the B-VID, so that the port MAC only needs to be installed in those transit nodes which lie on the shortest path between the service endpoints defined by the I-SID, and not on every node in the SPB region;

3. and collocated with a nodal MAC (we also consider this a useful property for some of the approaches considered for multiarea).

In general, I-SIDs and port MACs will be associated with a single B-VID only. Multiple B-VIDs are however allowed, because in some applications the number of I-SIDs may be sufficiently small that good load spreading cannot be achieved by assigning I-SIDs in this way. To overcome this, it is possible to spread a single I-SID across multiple equal cost trees, each defined by a VID, and to accept the penalty in forwarding state due to the per I-SID multicast state appearing in multiple VIDs. Then, traffic within a single I-SID could be spread over the multiple VIDs, using a technique like the hashing of C-MAC addresses or other header information deeper in the frame as used in Link Aggregation.

To minimize the impact of this volume of per service state on the network, several things must be done:

1. The state installed is the complete forwarding table for **nodal** MACs, so full connectivity is available to provide a useful level of basic OAM troubleshooting capability. This unicast state scales linearly (only # nodes x # VIDs), and so makes

installation of complete unicast connectivity by default a very modest burden.

2. **Port** MACs are only installed in the shortest path on the specified B-VID(s) between nodes with services in common. This is exactly the same "as needed" philosophy which governs multicast address installation, as outlined in the earlier introduction to SPBM ("Per-Service-Instance Routing and Forwarding," p. 24).

It is important to understand that the identity of nodal components, just like the naming of per service multicast trees, does not figure in the actual Dijkstra computation; the routes to them can be and are computed entirely by reference to the nodal MAC identity of the bridge advertising them. These finer granularity information tokens are only only processed at FDB generation time. Thus the computational scalability of SPBM is not significantly impacted.

Recasting the Group Multicast MAC Address for Shortest Path Trees

SPBV separates (S,G) by encoding the source of a tree in the SPVID, and a common group MAC address for all participants in the multicast group. In contrast, the group MAC address format used for the SPBM (S,G) trees encodes both the root of the multicast tree and the specific multicast group. Both the multicast bit, and the local addressing bit in the MAC address, are set to one. The 22 bit OUI is partitioned into two bits that define how the remainder is encoded (only 0b00 being specified at present), and the 20-bit nodal nickname of the root. The remaining 24 bits of the MAC address encode the I-SID.

An important implication of the (S,G) encoding is that multicast state scales as a square function (# services x # endpoints per service), in contrast to the linear scaling of unicast state. However, the impact of the square function for multicast is substantially mitigated by installation of multicast state on a "need to know" basis, in which only nodes lying on the SPF trees for a particular service install FDB entries for the addresses associated with that service on the ports actually used by the service.

Normally, in Ethernet a group multicast address requires no personalization when used by the set of senders. The encoding in the destination MAC corresponds to (*,G) and the combination of SA MAC and DA MAC provides (S,G) semantics. However, the SA MAC is not considered or needed for forwarding purposes, because the source is explicitly excluded from the set reachable from itself by the Ethernet split horizon forwarding model running over a spanning tree (a frame is never forwarded to the port on which it arrived). In SPBM, the group address is recast as referring to the set of receivers and connectivity specific to an individual source, but excluding the source itself. This means that the number of source-based trees per group (or (S,G) trees) goes up in proportion to the number of sources and services. This also means that the semantics of the group multicast address are subtly altered.

As Ethernet uses destination-based forwarding, a PBB-TE connection is actually defined in the data plane by the full 108 bits of MAC and VID information in the header. It can be conveniently decomposed into source interface, destination interface and forwarding plane instance identifier for the path between the two. The forwarding model constrains a bridge to look up only the DA and VID, meaning that traffic on a particular VID from all sources to a specific DA egresses by a single port, but the scaling is $O(N)$.

During standardization there was an interest in using the 802.1ah multicast encoding whereby $S = SA$, and $G = DA = \{802.1ah$ multicast $OUI + I\text{-}SID\}$. This was considered for backward compatibility reasons. However, the (S,G) instance then needs the full $SA + DA + ID$ (108 bits) for uniqueness, but the forwarding model would still only look up the $DA + VID$ (60 bits), and this would have required unacceptable changes to the Ethernet technology base.

An examination of other alternatives to keeping a common group address for all sources in a multicast group was not found to be fruitful. The use of a common multicast group address and reliance only on RPFC as a pruning mechanism means that indeed only one copy of a multicast frame reached a given leaf on the multicast tree, but in general roughly twice the required bandwidth was consumed in frames that RPFC discarded at intermediate nodes (see Fig. 2.5 and associated text for a brief explanation). This corresponds roughly to the basic RPF model of [Metcalfe]. It also has the problem of a greatly increased probability of generating looping scenarios due to the much less constrained distribution of superfluous multicast traffic.

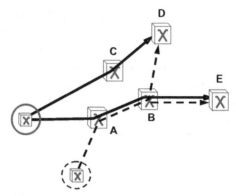

Figure 2.5 Different sources may not exhibit congruent trees.

Similarly when a more complex variation is considered, whereby a combination of RPFC and per port FDBs is treated as a replacement for a true (S,G, VID) 108-bit lookup, the result is still unsatisfactory. First blush suggests that aspects of downstream congruency resulting from the tie-breaking algorithm might be exploitable but this does not bear scrutiny.

The simple example in Figure 2.5 illustrates the problem. In this thought experiment, both solid- and pecked-ringed sources are using the **same** multicast, DA but the solid and pecked trees are not congruent at D and E. The result is that D will chronically receive multiple copies of frames from the solid tree, via C (as intended) and also from B (because the pecked tree requires that state to be installed at B). RPFC will cause those via B to be discarded at D, but this is still unacceptable from the point of view of waste of bandwidth.

Generalizing, the downstream congruency property is only true for the **intersection** of the set of nodes common to two SPF subtrees which transit a node, **not** their union. In other words, two SPF subtrees that transit a node (e.g., A in the figure) always use exactly same path to reach the downstream nodes which they have in common but, in general, the sets of downstream nodes in the two distinct SPF subtrees are not the same.

The diagram in Figure 2.6 illustrates the downstream congruency property where the SPF trees are identical for the subset of destinations they share but are not in themselves identical. This arises only because all three trees transit the node on the left.

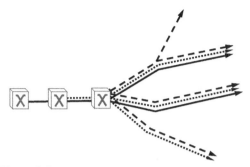

Figure 2.6 The condition for downstream congruency.

The belief that we could compress state and avoid full lookups via a combination of smaller lookups was a chronic mistake which we eventually trained ourselves to avoid.

Alignment with the original 802.1ah multicast encoding transpired not to be an absolute requirement of the SPBM effort. During the standardization process, the IEEE committee declined to consider a true 96-bit lookup as the only option, and warmed up to the use of the local bit with source nodal nickname and the I-SID for multicast DAs, providing the required (S,G) information compressed into the DA field.

ROUNDING OUT DESIGN DETAILS

In this chapter, we start to address specific topics in more detail. The first topic is a networking issue, OAM; then specific issues to be resolved by the control plane are addressed in the next chapters.

SPB OAM: Introduction

This is a very short section for a very complex subject. The reason is simple: SPB adopted the Ethernet forwarding, congruency, symmetry, addressing, and naming paradigm "as is," and accordingly could access the full Ethernet OAM suite developed originally for IEEE 802.1Q and IEEE 802.1ad. This comprises IEEE 802.1ag Fault Management functions, with Y.1731 for Performance Management.

OAM for SPBV is a direct reuse of 802.1ag and Y.1731 as originally envisioned for the Q-in-Q environment. Proactive fault management uses multicast addressing for OAM frames, and since MAC addresses can originate from the customer or the provider domain, the concept of maintenance levels is carried forward so that customer instrumentation of the portions of the network under their control and provider probing of their network can be properly separated.

These tools rely on eight well-known multicast addresses, each defining a different OAM "maintenance level" or Maintenance Entity Level (MEL), which are trapped and interpreted in accordance with the FDB at every bridge. The maintenance levels are typically configured to correspond to layering components in the overall Ethernet architecture. So, for example, one maintenance level could correspond to untagged MAC endpoints, another to C-tagged endpoints, and a third to S-tagged endpoints.

For an SPBV network, some maintenance levels will be reserved for use within the SPBV domain, while some will be delegated to the client layer and transparently relayed by the SPBV bridges. In this way customers of the network can verify their end-to-end connectivity without being able to probe details of the internals of the SPBV network.

For SPBM the B-MAC layer is not shared between the provider and the clients of the network. The visibility of unicast B-MAC addresses in the control plane allows unicast OAM, such as connectivity check messaging sessions, to be readily set up, either as needed for diagnosis, or continuously, as required for SPBM TE (see, e.g., "Topology Modification for Traffic Engineering Purposes," p. 183). While 802.1ag used well-known multicast addresses, both the IEEE and the ITU-T agreed that OAM frames received with other than the reserved addresses would be treated as "well formed." This foresight meant that the specified OAM could be extended into the 802.1ah realm, and by extension into 802.1aq also, with the receiver behavior unmodified and the sender utilizing appropriate 802.1aq MAC addressing for OAM messages.

Infrastructure OAM will likely be primarily confined to the role of diagnostic and adjacency liveliness detection. This means that the ping and trace-route paradigm must also be supported. It also requires that all BCB loopback addresses appear in the forwarding table so that link trace messages (LTMs) are not filtered by RPFC; for this reason, SPBM installs all nodal loopback B-MACs by default, irrespective of whether the disposition of I-SIDs indicates that a service will need them.

Service-level OAM is again expected to be primarily diagnostic, though since it runs entirely transparently above SPBM at the C-MAC layer, its use proactively ("always on") by a customer is not precluded.

The brevity of this section captures a significant post hoc realization; SPB, and PBB-TE before it, have been the first networking technologies to be developed where the OAM suite has been available, in full, **before** the control or management plane and the forwarding path have been integrated.

A useful item for future consideration would be applying IS-IS message exchange to configuration and distribution of relevant identifiers for OAM, following the precedent set by service membership distribution using I-SIDs.

SPB OAM: Summary of Tools

SPB directly inherits a comprehensive set of tools developed to instrument the Ethernet data plane. When we refer to OAM we are referring to the data plane OAM protocols that support network operations and fault/alarm and performance management, and not the craft and Element Management System (EMS) OAM interfaces that exist to support configuration and gathering of network and service statistics.

Ethernet OAM was originally forged in the IEEE 802.3ah project of the Ethernet working group, and was intended to address OAM for the first mile only, but later OAM focused on networking was developed jointly by the IEEE and ITU-T. Arising out of the PB project, there was a need to supply OAM for a number of functions.

A requirement on the OAM procedures for bridging was that they be solely dependent on the data plane, because the population of forwarding tables for bridging the flooding and learning mechanism operates in the data plane. This has several benefits: OAM works regardless of the control plane type, and even if a control plane is not employed, the OAM follows the true data path more closely, and it can be tunneled transparently over lower layers.

Ethernet has a number of architectural properties that make it amenable to the application of data plane OAM as a closed system. For example, bidirectional congruency is leveraged for loopback and fault management and performance management procedures.

The OAM functionality that has emerged from the IEEE and ITU-T includes a suite of fault management, alarm management, and performance monitoring tools. These are defined as a protocol suite and exercise a range of functionalities in the data path typically by using the same frame formats and forwarding procedures as normal traffic, with the OAM flows being distinguished from regular traffic by the receiver using the Ethertype.

The IEEE tools provide a basic fault management suite which includes:

CCM—the connectivity check message: it is a multicast heartbeat using a reserved address (Y.1731 defines the unicast heartbeat message);

ETH-LB and LT, loopback and link trace, analogous to IP's ping and traceroute.

The ITU-T Y.1731 tools both augment the fault management set by defining unicast variations of the fault management tools, and add performance monitoring transactions such as

LM—loss measurement

DM—delay measurement

Of lesser utility in a routed system, a set of alarm management transactions were also added by Y.1731:

AIS—alarm inhibit signal

RDI—reverse defect indication

Finally, Y.1731 defined a number of capabilities which provide an opaque container set for future applications:

MCC—maintenance communication channel

VSP—vendor specific extension

EXP—experimental extension

At the time of writing, the Metro Ethernet Forum was working with the ITU-T on service OAM extensions for performance management. The work was based on adding loss measurement transactions based on synthetic traffic. The motivation for this was to address the issue that in a true bridged environment, the amount of unicast traffic which had been flooded as unknown could not previously be distinguished from traffic forwarded entirely on learned paths, and hence an apples to apples comparison between the send counts at the network ingress could not previously be performed with those at the egress. Such extensions will be required for SPBV, and may provide value for SPBM simply due to the reduced state requirements of an on-demand test.

RPFC for Loop Mitigation

RPFC applied to Ethernet was a new suggestion, but now no longer really counts as an extension to Ethernet. The concept has been taken up by the IEEE 802.1aq project standardizing Shortest Path Bridging, and it is a minimum-impact re-purposing of mechanisms intrinsic to bridging since its inception. In a learning bridge, every ingress frame must be inspected in case its source MAC is unknown. If it is unknown, the learning procedure must be invoked to place that MAC into the FDB against that ingress port. RPFC merely requires that same invocation on unknown source, but now to drop the frame, which is a much simpler operation than the original.

CHAPTER 3

Why the SPB Control Plane Looks as It Does

THE CONTROL PLANE IS AS SIMPLE AS IT CAN BE, BUT NO SIMPLER

This chapter explains how essentially all the control functionality for SPB can be delivered by the routing system alone. We start with a brief introduction to the task to be performed and introduce the concept of putting all the required information to construct all aspects of data plane forwarding into the routing system. We then discuss in some detail SPBM's most radical departure from previous received wisdom, the complete elimination of signaling from both unicast and multicast state installations. We then provide a factual introduction to the extensions to the Intermediate System to Intermediate System (IS-IS) routing protocol required by SPB, showing how modest these are. Finally, we explain some of the algorithmic innovations over previous link state practices that are exploited by SPB.

The Layman's Description of What the Control Plane Does

SPB is primarily computation driven. Although the full details of how the SPB computation is best performed can be subtle and intricate, what

802.1aq Shortest Path Bridging Design and Evolution: The Architect's Perspective,
First Edition. David Allan and Nigel Bragg.
© 2012 the Institute of Electrical and Electronics Engineers. Published 2012 by John
Wiley & Sons, Inc.

it does is relatively easy to describe. Given the current network state that has been advertised and synchronized across the network by IS-IS, a node will compute the set of shortest paths that go through it and, for each of these paths, the associated pair of endpoint nodes. It will then determine the intersection of the sets of VLANs (SPBV) or I-SIDs (SPBM) associated with each of these pairs of nodes and then populate the local filtering database (FDB) with both the unicast addresses of the endpoint nodes and the shortest path VID (SPVIDs) in the case of SPBV or algorithmically constructed per service multicast addresses for SPBM. In the case of SPBM, each per service multicast tree is further customized according to the transmit/receive attributes of the intersection set. Although SPBV does not have this concept of "service," it also allows the default broadcast tree for a VLAN to be pruned by MVRP registrations of interest in a VLAN at the edge, and internally, the same transmit/receive attributes are used to convey this. When all the nodes in the network have completed this operation, the required unicast and multicast connectivity, per service in the case of SPBM, will exist.

SPB uses the Dijkstra algorithm for shortest path computation. For those who have not encountered the Dijkstra algorithm before, there is an excellent visual analogy found in [SPT- HTzeng]:

- Represent each node in the network by a ball bearing.
- Represent each link by a length of thread with a length proportional to link cost ("better" = shorter thread) and with each end tied to the appropriate bearing.
- Pick up the "root" node bearing; let the others hang under gravity.
- The tight threads define the shortest path(s) from the root to all the other nodes in the network.

Enhanced Role of the Routing System

IS-IS was chosen as the preferred link state protocol, primarily for its layer 3 independence.

The requirement for SPBM to construct per service (I-SID) multicast trees meant that IS-IS needed to be augmented to carry per Backbone Edge Bridge (BEB) service information. As I-SIDs have domain-wide significance and require no personalization, it was natural to simply reuse them without modification in the control plane and to directly correlate control plane and data plane behavior.

Moving from computing only the shortest path to each BEB to the generation of congruent unicast and multicast trees required additional computation on the part of the routing system. The computation output permitted each node to determine the set of node pairs for which it was on the shortest path between the members of the pair. Furthermore, the node performing the computation could also determine if and when members of a pair had service instances (Base VIDs for SPBV and I-SIDs for SPBM) in common and install the requisite connectivity.

Using the "ball and thread" analogy introduced above, to compute "all-pairs shortest path" as required by SPB, each node conceptually carries out the process outlined earlier for every bearing in turn, determining for each whether the computing node lies on a tight thread, and if so, which bearings hang on tight threads from beneath it.

Floyd's all-pairs algorithm was considered as an alternative to computing multiple Dijkstras, one per node in the network. Ultimately, the multiple Dijkstra approach prevailed due to its superior performance, and it was used as the baseline approach for subsequent algorithm improvements. In essence, Dijkstra at any given time maintains state in proportion to the circumference of the circle defined by the current distance from the root under consideration, rather than the area of the circle, and this state reduction gives it a computational edge as the processing effort spent maintaining and traversing the state is diminished accordingly.

Elimination of Signaling from Multicast Tree Setup

In terms of revisiting received wisdom, SPBM's most radical innovation is the use of global addressing and all-pairs shortest path to eliminate the need for signaling from multicast tree computation. [M-OSPF] had already explored this territory in an era when it was, in practice, found to be computationally intractable, and hence the application of conventional approaches has continued unabated until the present day. To allow the reader to gauge the significance of this, this section is written as an explicit comparison between SPBM and the most modern "conventional" multicast protocols supporting MPLS multicast. The single example of MPLS is used for ease of exposition, but the comparison is general in its fundamentals since the MPLS protocols are a direct rework of older IPs such as PIM-SM.

SIDEBAR: *The Rise of Computing Power*

SPB's elimination of signaling by the use of all-pairs shortest path is computationally intensive, and it is worth placing this load in the context of the advances in computing power since link state protocols were first deployed.

The first [IS-IS] specification was published in 1992, and the protocol was deployed in the early days of the Internet in the mid-1990s. The solid line in Figure 3.1 shows the two or more orders of magnitude increase in semiconductor process capability since that time, as measured by transistors per device (source: Wiki/Intel). Furthermore, the raw processor clock rate has also increased by approximately an order of magnitude during this same period.

Thus, whereas a single Dijkstra may have consumed the available computer cycles for route computation in the mid-1990s, the all-pairs shortest path is now computationally tractable for substantial network sizes. Indeed, the practical bottleneck in restoration times is now generally found to be the download of forwarding tables to line cards, not the tree computation process itself.

Finally, we observe that as process technology approaches limits in terms of feature size and clock rate, the modern design direction is the use of multicore processors. The all-pairs shortest path computation requires many independent tree calculations, of the order of one per bridge in the network, to be performed on the same topology database. Consequently, this computation is ideally suited to the kind of coarse-grained parallelism supported by multicore designs.

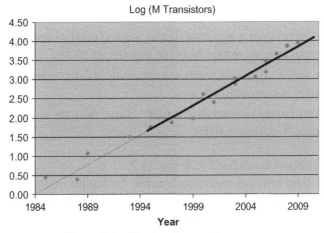

Figure 3.1   The rise of computing power.

Multicast Mechanisms

Both multicast LDP (mLDP; M-LDP) and SPBM produce p2mp multicast trees with similar properties, shortest path trees (SPTs) with a single path to the root that tend toward a minimum-cost solution. This is a consequence of the equal-cost tie-breaking techniques used in each. The mLDP protocol uses the opaque value in the FEC which identifies the multicast group as an input into equal-cost tiebreaking. A simple modulo computation using a given opaque value will cause nodes presented with the same set of next hops to resolve the same way, so minimizing hop diversity for any single tree. SPBM uses an edge-to-edge path tie-breaking algorithm that produces minimal-cost SPTs (minimum hop diversity) but goes further in that the multicast trees produced are both congruent with unicast forwarding and symmetric between any two points. A degree of path diversity between each multicast tree can be achieved by mLDP; SPBM offers path diversity between different B-VIDs, where each B-VID defines one distinct full mesh of connectivity across the entire network.

The mLDP protocol uses a receiver-initiated join paradigm, with no inherent mechanism to advertise the set of senders. This is out of scope and left to some other protocol. So, the addition of a node to an (*,G) multicast group involves the node learning the set of existing sources $\{S_i\}$ = * in (*,G), and then signaling receiver interest in each source in (*,G), and in addition, all peers in the set of (*,G) must learn of the new node and signal their receiver interest to construct a tree rooted on the new addition. This is a not insignificant number of transactions as

- the addition of a new root or leaf is advertised to each root (order N transactions, where N is the number of roots);
- the new root or leaf needs to initiate "joins" to each existing root (order N transactions); and
- each existing root needs to initiate "joins" with the new root or leaf (order N transactions).

The process is outlined in Figure 3.2.

This transactional tree construction technique uses the "downstream unsolicited, ordered mode, conservative label retention" form of LDP label distribution, in which a node only computes the parent in

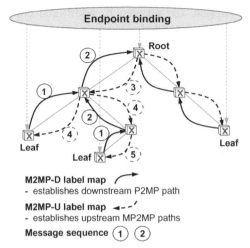

Figure 3.2   Signaling to support multipoint-to-multipoint tree setup.

a given tree and depends on signaling from children to learn that there are children, which inherently results in a protracted hop-by-hop convergence process.

SPBM advertises both the set of senders and the set of receivers in IS-IS for a given group (advertised as an 802.1ah I-SID, the layer 2 service identifier, augmented with send/receive attributes). The act of a node flooding a single IS-IS LSP advertising its interest in a service instance and its associated send/receive interest is sufficient information for the network to add the node as a leaf to those specific trees it is interested in receiving from, and adding the other nodes as leaves to a service specific tree originating at the advertising node. So, while mLDP learns the topology, computes the shortest path to the root, and then initiates hop-by-hop signaling to construct the multicast distribution tree (MDT), SPBM learns via IS-IS flooding, and then the individual nodes compute the MDT directly. This is a substantially faster process, which is of particular relevance to rapid restoration.

Loop mitigation is significantly different between the two. As mentioned, mLDP uses break before make (because a node only holds one upstream label binding at a time for a given label switched path, with directed trees (where possible), and with time-to-live (TTL) expiry as loop mitigation mechanisms. This typically means that while frame duplication can happen, the number of additional copies received by group members will be bounded.

As was discussed earlier, SPBM uses control plane handshaking (in the form of synchronization determined by the exchange of database digests) as its fundamental loop avoidance mechanism. As a reminder, whenever a topology change occurs, then the multicast forwarding state is removed for those trees where the cost to the root has changed. New state is only reinstalled when peers agree on the view of the topology upon which they have each locally converged, which by implication means they have reestablished agreement on the distance to all roots.

This approach has a number of desirable properties. First, we maintain uninterrupted connectivity for multicast trees unaffected by the topology change. Second, synchronization of multicast updates does not need to be ordered from the root; nodes can safely reinstall the affected state as soon as they are synchronized with relevant peers because if a peer has not achieved the required synchronization further up the tree, its own lack of installed multicast state "protects" the downstream nodes. Finally, the delay that synchronization would normally be expected to incur is largely eliminated, as the required handshaking with peer nodes can be done in parallel with the computation of the multicast FDB.

This technique is sufficiently robust for multicast loop avoidance to even permit the application of multicast to the control plane itself, to eliminate delays in the propagation of topology changes, and to accelerate network convergence.

SIDEBAR: *Tree Construction Styles*

When considering paths and trees, there are several styles that can be considered. The differences determine the relative suitability of these depending on the application of the tree to unicast and/or multicast forwarding.

For example, if neither unicast efficiency nor latency is a consideration, the ideal multicast-only tree will simply traverse a minimum number of hops (see Fig. 3.3). But it is easy to see that the number of bandwidth hops consumed by such a tree for unicast is not necessarily particularly efficient or desirable.

Similarly, if multicast is not a consideration and unicast efficiency is the only measure of importance, an SPT that spread the traffic as much as possible would be desirable, as shown in Figure 3.4.

If state is the primary consideration, a common spanning tree is the most efficient, as one tree serves the entire set of sources. There is no set of per source trees. SPB can compute such common spanning trees in a straightforward manner without significant change to the core algorithms. Each service instance (I-SID) is associated with a single root bridge, and (*,G) multicast state for that service instance is installed on the SPT built on that bridge and pruned to match the I-SID endpoints.

The default behavior for SPB is to construct minimum-cost SPTs that offer a reasonable compromise between unicast and multicast bandwidth efficiency while preserving the unicast–multicast congruence property. This is a consequence of the tie-breaking algorithm. What this means is that, when a given set of child nodes has an overlapping set of equal-cost paths to a given root, all children will choose the same parent as illustrated in Figure 3.5. It is easy to envision that the net result of this is a minimum-cost SPT simply because the number of transit nodes in such a tree will be minimized.

SPBM can support multiple tree construction algorithms simultaneously, using the B-VID as the data plane separator, and the mapping of services to B-VID defines how applications would be associated with optimized trees. The same capability can be supported by SPBV using multiple SPVID sets.

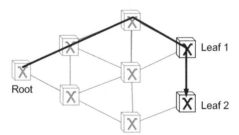

Figure 3.3 Minimum-cost MDT.

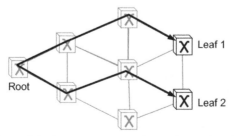

Figure 3.4 Shortest path tree with maximum diversity.

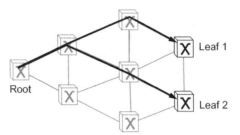

Figure 3.5   Minimum-cost shortest path tree.

Convergence Behaviors

In the absence of resiliency helpers such as fast reroute (FRR), considered below and shown not to be helpful, mLDP's combination of break before make and signaling after convergence tends to suggest relatively slow recovery from failures. SPB's use of in-place computation, with no need for signaling, and its application of multicast to the control plane achieves much faster overall restoration times in metro-sized networks without the need of "helpers," as we now discuss.

mLDP operates in a form analogous to "ordered mode with conservative label retention" such that only currently active label bindings are retained between peers, and the existence of a binding is overloaded to actually define the forwarding path. Hence, a downstream label binding for a particular label switched path is required for forwarding. When an mLDP system is informed of a topology change, it must first compute a new SPF solution, and then invalidate any multicast labels previously offered to upstream peers that have been rendered obsolete and notify those upstream peers of this, and then finally offer new labels to the new upstream peers.

The ordered mode aspects of this have undesirable side effects during the transition of a node from a transit role to one no longer participating in a given multicast label switched path. This produces a number of potential race conditions, especially for sparse trees, which are an illustration of the emergent behaviors of this set of tightly coupled but asynchronous protocols. A transit node will continue to try to maintain a label binding with a parent for a given multicast path as long as it believes it has children, manifested by label bindings, on that same path downstream of itself. An example scenario would be a transit node that will cease to be a transit node after the network has converged. If it converges before the child nodes, it will initiate removal of the original upstream binding and may get as far as establishing a new upstream binding prior to the child nodes converging. When it later discovers that the child node has removed the label binding because it has just found a new shortest path, the now ex-transit node must similarly remove the binding it just offered to the parent. Since the interaction of signaling and control plane computing capacity is the rate-limiting step in convergence, heavy messaging loads repeatedly causing the control plane to suspend useful route computation means that gratuitous behavior of this sort will be directly inimical to overall network performance.

SIDEBAR: *Just How Many Messages Does It Take to Converge a Reasonable Network?*

This is a simple example to illustrate computational convergence compared with transactional convergence. In the example illustrated in Figure 3.6, we show a pair of five-connected nodes "A" & "C," where A is a "hot spot" for 5000 p2mp MDTs, which might be, for example, 1000 "five-site" E-LAN services:

- ~1000 "roots" downstream (behind B)
- ~4000 upstream (across the MAN)

If we fail the link from A to B, the implication is that

- the path from "A" to the 1000 roots behind "B" is severed;
- the path from B to the 4000 other roots in the MAN is affected;

(Continued)

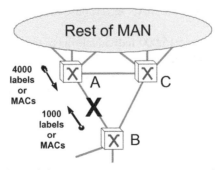

Figure 3.6   A multicast reconvergence scenario.

- the MAN will switch transit for B and beyond to go via C instead of via A.

Tie-breaking procedures in both mLDP and SPBM mean that, in a symmetrical network, MDTs traversing A will not likely traverse C under fault-free conditions, since both produce minimum-cost SPTs in the presence of equal cost.

So, in an mLDP environment subject to the fault shown in Figure 3.6, **A**

- advertises the failure in the IGP to all peers   5 LSPs
- withdraws 4000 labels offered to the MAN   4000 transactions
  (from receivers behind B binding to sources across the MAN)

and **B**

- advertises the failure in the IGP to all peers   4 LSPs
- offers 4000 labels to C to reroute around A   4000 transactions
  (receivers behind B binding to sources across the MAN)

and the **MAN**

- withdraws 1000 labels from A   1000 transactions
- offers 1000 labels to C   1000 transactions
  (groups across the MAN binding to sources behind B)

and **C**

- offers 1000 labels to B in response to MAN   1000 transactions
  (groups across MAN binding to sources behind B)
- offers 4000 labels to the MAN in response to B   4000 transactions

This is a total of some 15,000 individual transactions, which, if well synchronized in time, could be packed into 300 or so signaling messages.

SPBM operation eliminates the signaling. A and B both advertise the failure by flooding LSPs to their peers, and these are the only transactions required.

SPBM does not require a signaling system for multicast and is able to operate in a mode close to what MPLS would term "independent mode," but which is actually an aggregated ordered mode because of the interaction of loop avoidance with the convergence process. SPBM substitutes computation for signaling such that transit nodes learn when they are on a given multicast tree as an integral part of FDB generation. When an SPBM node has performed the computations on a topology database, it knows when it is on the shortest path between a root and one or more leaves and can install state accordingly. Convergence is not gated by incremental discovery via signaling transactions of a node's place on any individual multicast tree. Where SPBM does differ from independent mode is its enforcement of agreement on the current network topology with its peers, which was referred to earlier as an "aggregated ordered mode." This very efficient mechanism uses exchange of a single digest of link state covering the entire network view and does not need agreement on each path to each root individually, which is very costly in state and transaction volume. This is inherent to the loop avoidance procedures in the control plane. The result is that the volume of messaging exchanged to converge the network is in proportion to the incremental change in topology and not to the number of MDTs in the network.

The mLDP protocol (and, for that matter, multicast in general) cannot benefit from resiliency "helpers" such as FRR in the same way as can LDP for unicast connectionless traffic, or as indeed can multicast replicated at ingress onto a "split horizon" unicast core. FRR's primary utility is as a helper for connectionless traffic whereby there is a complete forwarding table (LFIB) for every destination in all nodes, and all nodes promiscuously accept traffic from all peers. The result in this scenario is that there are intermediate states in the network where connectivity, albeit in a nonoptimal form, will exist throughout any periods of transient instability while the network reoptimizes itself. However, the p2mp directed trees employed by mLDP (which only offer a single label binding at any time), and SPB's directed trees, cannot really benefit from techniques such as FRR because as soon as the network begins to react and reoptimize itself, the connectivity will again be disrupted, with no further recourse to maintain connectivity until the network has converged. FRR cannot significantly help the restoration of multicast connections, and multicast is really a connection-oriented mode of operation. Connection restoration requires further helpers such as the "make before break" procedure of RSVP-TE.

It is worth a few words on the feasibility of applying SPBM multicast techniques to MPLS. The summary conclusion is that the 20-bit label size is inadequate. If partitioned such that absolute multicast labels can be algorithmically constructed, then the scalability is perhaps comparable to Q-in-Q; postulating 1 bit as a multicast indicator and 7 bits for the nodal ID leaves 12 bits for the service ID, precisely the service space supported by Q-in-Q. The alternative route to rapid restoration would be the recreation of "liberal label retention" for multicast, which restores the efficiency of utilization in the label space, but the state in the control plane would then scale according to the product of the number of services × end-points per service × adjacencies per transit node.

SIDEBAR: *Liberal Label Retention*

Liberal label retention is how MPLS unicast minimizes the number of transactions to be exchanged after a network topology change by actually predistributing all state against all alternative routes that might be needed in advance. What it does is to drive the state up in proportion to the number of adjacencies a given node has, as a full set of inactive bindings maintained in addition to the currently active set. When combined with per-platform labels, some simplifications can be realized as the information offered to all neighbors is the same.

Unstructured liberal label retention applied to multicast would make much more efficient use of the label space compared with the use of algorithmically constructed global labels. This is simply because imposing structure on the label introduces inefficiency in any sparse distribution of services as a significant portion of the label space becomes effectively "stranded." However, "liberal" will still drive state up, as a given node will still a priori have a label binding per peer interface per service per root, compared with a service binding to a root. As with unicast, it inflates the amount of state in the control plane in proportion to the number of interfaces.

The properties of SPBM allowed us to move down a different path and to exploit the insight that the greatest possible dividend is found by directly minimizing convergence times and that the primary barrier to convergence is internodal per tree signaling. In short, SPBM addresses and accelerates complete unicast and multicast convergence by leveraging multicast itself for the distribution of IS-IS LSPs. The mechanisms used for this are described later in "Fast Fault Notification" (p. 103).

Ultimately, SPBM requires significantly less state and less state exchange, and consequently will converge multicast forwarding much faster than mLDP.

The principal reason for this is that by its nature, mLDP pretty much **has** to replicate the traditional multicast paradigm, first computing the unicast topology and only **then** signaling multicast interest. This has been replicated throughout the industry (be it Spanning Tree Protocol and MMRP, IGMP, PIM-SM, etc.). SPBM renders this model obsolete by applying modern levels of computing power to multicast FDB generation.

CONTROL PLANE INFORMATION

IS-IS: A Thumbnail Sketch

Before discussing the information that is carried in the SPB control plane, we first provide a very brief introduction to IS-IS so that those not familiar with the protocol have some context with which to understand how information is handled within it. Disclaimer: this makes no attempt to be an exhaustive description; it merely summarizes the key principles and behaviors.

IS-IS is a link state routing protocol originally designed to route the ISO connectionless network protocol and only later extended to route IP (RFC 1195 et seq). It is thus independent of IP and runs directly over layer 2. It supports strictly hierarchical multiarea routing (known as "levels" in IS-IS). The overall network design issues for SPB which this raises are discussed later ("Multiarea," p. 171).

Information exchanged between IS-IS routers is carried in a small number of classes of protocol data units (PDUs):

- IS-IS Hello packets (IIH),
- Link State Packets (LSPs),
- complete sequence number packets (CSNPs), and
- partial sequence number packets (PSNPs).

There are three types of Hello packets: point to point, broadcast medium level 1, and broadcast medium level 2. The others have two types, for level 1 and level 2. This use of different PDU types for different levels avoids all ambiguity. The PDU types at the different levels have the same basic function, but there can be subtle differences in the parameters that may be carried.

Hello packets are exchanged between adjacent routers and are not propagated further. Their role is fundamentally to establish that compatibility exists to allow an adjacency to exist.

The PDU has a fixed format header defining the protocol and version, addressing limits, the system (source) identifier, nodal holding time value, and so on. This may be followed by other parameters, called "variable length fields" in IS-IS, but often known as type-length-value (TLV) coded parameters elsewhere. These advertise the set of Area Addresses of this router, the system identifiers of immediate adjacencies on this LAN, and the authentication parameters of this router.

The function of **LSPs** is to flood information throughout the IS-IS domain. The header contains almost the same information as the Hello packet, with the addition of a remaining lifetime and a sequence number. These latter pairs of parameters are key to IS-IS reliable flooding. Originators of LSPs stamp them with a validity period and must reissue them with an updated sequence number before their validity expires. Stale LSPs are thus aged out even if a failure blocks all messages invalidating them explicitly. Recipients of an LSP flood them to all adjacencies, apart from back to the one from which they received it, but only if they have not received that sequence number before. In this way, each LSP is flooded onto every link between adjacencies but, at most, only once in each direction.

Following the LSP header, there are again parameters that are each carried in a TLV:

- The source identifier of an immediate neighbor (as carried in the Hello packet) is extended in this PDU type to include the cost metric of the associated link.

- The address of an endpoint (in level 1) or reachable prefixes (in level 2)

CSNPs and PSNPs are exchanged between adjacencies as part of the process of maintaining and guaranteeing link state database synchronization:

- A CSNP describes all known LSPs in the link state database. Each TLV in the packet includes the following information about an LSP:

- its source identifier, its remaining lifetime, and its sequence number.
- PSNPs are similarly constructed to CSNPs but have summaries of specific LSPs only. They are used in two ways:
  - to acknowledge receipt of an LSP on point-to-point links, and
  - to request transmission of the latest version of an LSP.

Visual Model of Control Plane Information

The previous section introduced IS-IS. This section introduces the new information items needed for IS-IS for SPB. These information elements are documented in [IS-IS-L2] and [IS-IS-SPB].

The new items associated with a node are modest in number. Referring to Figure 3.7, the nodal nickname is the 20-bit value used to construct the service-specific (S,G) multicast addresses. SPB has its "own" link metric field to avoid any interaction with other IS-IS applications. The digests are a compact topology summary, used to determine whether or not nodes share an identical topology view, which is a key part of the synchronization process used to guarantee loop-free forwarding at all times.

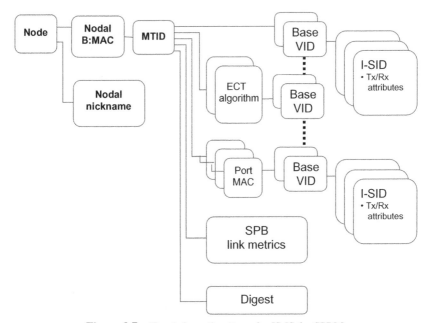

Figure 3.7 New information items for IS-IS for SPBM.

The remaining items are the nodal unicast media access control (MAC) addresses, B-VIDs, and the services associated with each. The nodal B-MAC is the SPB equivalent of the IP loopback address and need be the only externally visible address in an SPB domain. However, as discussed earlier ("Meaning of MAC Addresses in PBT and SPBM; Port and Nodal MACs," p. 64), PBB permitted different granularities of B-MACs to allow implementation trade-offs to be made. Multiple B-VIDs for ECMT may be associated with SPBM operation, and hence the B-MACs and services (I-SIDs) are associated with (usually) a single B-VID. I-SIDs can be associated with more than one B-VID, for example, in enterprise applications with small numbers of I-SIDs, where a single service can be load balanced across multiple B-VIDs.

The equivalent information items for SPBV are shown in Figure 3.8. The only new item compared with SPBM is the SPVID, the unique nodal virtual LAN identifier (VID) used by a bridge to identify its SPT in the forwarding plane of a particular VLAN. Because SPBV uses MAC learning, the only visible MAC is the "loopback" MAC of the bridge, used as part of its identifier in IS-IS.

New IS-IS TLVs for Link State Bridging

All the new IS-IS TLVs are described briefly in this section. Because the details of their structure are still being refined in the Standards bodies, we felt it inappropriate to give what could only be a snapshot

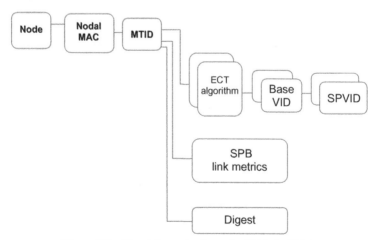

Figure 3.8 New information items for IS-IS for SPBV.

of a work in progress; instead, we provide an informal summary of the key information elements in each and their significance. The definitive version of this material may be found in [IS-IS-L2] and [IS-IS-SPB].

Multitopology is introduced as well, but multiple instances of SPTs for SPB can be described in a single topology instance defining multiple Base VIDs. Multiple topology instances allow different topologies in the future if it is desired to use different metric sets since there is only a single metric scheme allowed within each topology.

Link state bridging introduces no new PDUs to IS-IS and adds only new TLVs and sub-TLVs to the existing IS-IS PDUs:

(a) ***Multitopology Aware Port Capability TLV (MT-PORT-CAP).*** This differentiates topology instances in Hello (IIH) PDUs:

- This carries an MT identifier (for possible use in the future— see above)

- and an overload bit specifically for use by link state bridging to indicate whether the bridge can be used for transit, following the analogy of the generic IS-IS overload bit.

(b) ***SPB MCID Sub-TLV.*** This sub-TLV is added to an IIH PDU to communicate the digest for multiple spanning tree configuration identifiers (MCIDs). This digest is used to determine when the bridges' Ethernet configurations are exactly matched, which is a precondition for forming an IS-IS-SPB adjacency. The data used to generate the MCID are populated by configuration and are a digest of the VIDs allocated to the various Ethernet protocols. Two MCIDs are carried to allow transitions between different but nonconflicting configurations, by permitting an IS-IS-SPB adjacency to remain "up" as long as one MCID of the pair advertised by each bridge matches one of the MCIDs advertised by its neighbor.

The information elements are

- *MCID (50 Bytes).* The complete MCID as defined in IEEE 802.1Q. The set of interconnected shortest path bridges with identical MCID values identifies an SPT region.

- *Aux MCID (50 Bytes).* The complete MCID defined in IEEE 802.1Q, which identifies an SPT region. The aux MCID allows SPT regions to be migrated by the allocation of new VLAN to FDB mappings.

The SPB MCID sub-TLV is carried within the MT-PORT-CAP TLV, and this is carried in an IIH PDU.

(c) ***SPB Digest Sub-TLV.*** This TLV is added to an IIH PDU to indicate the current topology digest value. This information should converge to be the same on all bridges in an unchanging topology. Matching digests indicate (with extremely high probability) that the topology view between two bridges is synchronized, and this is used to control the updating of forwarding information. Digest construction is considered later in the control plane description, under "Agreement Digest Construction Details" (p. 115).

During the propagation of LSPs, the agreement digest may vary between neighbors until the key topology information in the LSPs has converged to become the same. The digest is therefore a summarized means of determining agreement on database consistency between nodes and may hence be used to infer agreement on the distance to all multicast roots. The digest TLV contains the following information:

- A (2 bits) The *Agreement Number* (*AN*) 0–3, which aligns with the AN concept of [SPB], used to guard against control packet loss or out-of-order delivery. When the agreement digest for this node changes this number is incremented. The node then checks for agreement digest match (as below). The new *local AN* and the updated *local Discarded AN* are then transmitted with the new agreement digest to the node's neighbors in the Hello PDU. Once a *local AN* has been sent, it is considered outstanding until a matching or more recent *Discarded AN* is received from the neighbor.

- D (2 bits) The *Discarded AN* 0–3, which aligns with the AN concept of [SPB]. When an agreement digest is received from a neighbor, this number is set equal to the *received AN* to signify that this node has received this new agreement and has discarded any previous ones. Then,

IF the local and received agreement digests match,

  THEN *local Discarded AN = received AN* + 1

  IF *received Discarded AN = = local AN* + **N** (where **N** = 0 or 1)

  THEN the node's topology matches its neighbor.

Whenever the *local Discarded AN* relating to a neighbor changes, the local agreement digest, *local AN*, and *local Discarded AN* are transmitted.

- The Agreement Digest. This digest is used to determine when IS-IS is synchronized between neighbors. The agreement digest is a hash computed over the set of all SPB adjacencies (all edges) in all SPB multitopology instances. In other words, the digest includes all VIDs and all adjacencies for all MT instances of SPB. This reflects the fact that all SPB nodes in a region must have identical VID allocations, and so all SPB MT instances will contain the same set of nodes. The procedure for computing the agreement digest is given later ("Agreement Digest Construction Details," p. 115).

The SPB Digest sub-TLV is carried within the MT-Port-Capability TLV, which in turn is carried in an IIH (Hello) PDU.

(d) *Multitopology Aware Capability TLV.* Differentiates topology instances for other SPB TLVs in LSPs.

(e) *SPB Base VLAN Identifiers Sub-TLV.* This sub-TLV is added to an IIH PDU to indicate the mappings between ECT algorithms and Base VIDs that have been configured on the advertising bridge. This information should be the same on all bridges, and this is verified by the digest carried in the SPB MCID sub-TLV described above. It is the values carried in the SPB MCID sub-TLV that determine whether an IS-IS-SPB adjacency can be formed or maintained, not the explicit mappings carried in this TLV. Discrepancies between neighbors with respect to this sub-TLV are temporarily allowed during upgrades (e.g., during the assignment of new ECT algorithms to Base VIDs), but all active Base VIDs, as declared by the state of the Use-flag below, must agree and use the same ECT algorithm. The key information element is a list of ECT-VID tuples, each comprising

- ECT Algorithm (4 bytes). The ECT algorithm is advertised when the bridge supports a given ECT algorithm (by OUI/Index) on a given Base VID. There are 17 predefined IEEE algorithms with index values 0–16 and the IEEE OUI occupying the top 24 bits of the ECT algorithm.
- Base VID (12 bits). The Base VID that is associated with the SPT set.

- Use-Flag (1 bit). The Use-flag is set if this bridge, or any bridge that this bridge sees is currently using this ECT algorithm and Base VID (as determined by the state of remote U-bits in the SPB Instance sub-TLV).

- M-Bit (1 bit). The M-bit indicates if this is SPBM or SPBV mode.

The SPB base VLAN identifier sub-TLV is carried within the MT-PORT-CAP TLV, and this is carried in an IIH PDU.

(f) *SPB Instance Sub-TLV.* The purpose of this sub-TLV is to flood throughout the SPB domain the information about the advertising bridge that all other bridges in the domain must know in order to construct the common nodal view of the domain. The SPB Instance sub-TLV gives the SPSourceID for this node/topology instance. This is the 20-bit value used for formation of multicast DA addresses for frames originating from this node and topology instance. The SPSourceID occupies 20 of the upper 24 bits of the multicast DA, following the group and locally (administered) bits, and two bits reserved for indication of future SPSourceID assignment modes. This sub-TLV is carried within the MT-Capability TLV in the fragment zero LSP.

The information elements comprise

- CIST Root Identifier (64 bits). The CIST root identifier is for interworking with RSTP and MSTP at SPT region boundaries. This is an imported value from a spanning tree.

- CIST External Root Path Cost (32 bits). The CIST external root path cost is the cost to the root of the tree as computed by the spanning tree algorithm.

- Bridge Priority (16 bits). Bridge priority is the 16 bits that, together with the low 6 bytes of the IS-IS System ID, form the spanning tree compatible bridge identifier. This is configured exactly as specified in [802.1Q]. This allows SPB to build a compatible spanning tree using link state by combining the bridge priority and the IS-IS System ID to form the 8-byte bridge identifier. The 8-byte bridge identifier is also the input to the 16 predefined ECT tie-breaker algorithms.

- V-bit (1 bit). The V-bit (SPBM) indicates this SPSourceID is autoallocated. If the V-bit is clear the SPSourceID has been configured and must be unique. Allocation of SPSourceID is defined in [SPB]. Bridges running SPBM will allocate an SPSourceID if they are not configured with an explicit SPSourceID. The V-bit allows neighbor bridges to determine if the auto allocation was enabled. In the rare chance of a collision of SPsourceID allocation, the bridge with the highest priority bridge identifier will win conflicts and the lower priority bridge will be reallocated, or if the lower priority bridge is configured, it will not be allowed to join the SPT region.

- The SPSourceID is a 20-bit value used to construct multicast MAC DAs for frames originating from the originating node of the LSP that contains this TLV.

- A List of (ECT Algorithm, Base VID, plus Flags) Tuples. Each ECT algorithm is associated with a Base VID, an SPVID for SPBV, and some flags described next. Each ECT-VID tuple comprises the information given earlier, under the SPB base VLAN identifiers sub-TLV, with the following additions:

  - U-Bit (1 bit). The U-bit is set if this bridge is currently using this ECT algorithm for I-SIDs, which it itself sources or sinks. This is a strictly local indication; the semantics differ from the Use-flag found in the Hello, which will set the Use-Flag if it sees other nodal U-bits are set or it sources or sinks itself.

  - M-Bit (1 bit). The M-bit indicates if this algorithm is being used in SPBM mode (when set) or SPBV mode (when clear).

  - A-Bit (1 bit). The A-bit (SPBV) when set declares that this is an SPVID with autoallocation. Since SPVIDs are allocated from a small pool of resources (typically 1000 or less), the chances of collision are high. To allow autoallocation, LSPs are exchanged with the bridge requiring an allocation setting its SPVID to 0, and then setting its SPVID to the operational value once it has bid for and obtained its allocated space. The SPVID value may also be configured.

(g) *SPB Instance Opaque ECT Algorithm Sub-TLV.*

(h) *SPB Adjacency Opaque ECT-ALGORITHM Sub-TLV.* There are multiple ECT algorithms already defined for SPB; however, additional algorithms may be defined in the future. These algorithms will use this optional TLV to define tie-breaking data for the new algorithm. There are two broad classes of algorithm: one that uses nodal data to break ties and one that uses link data to break ties, and so as a result, two identically formatted TLVs are defined to associate opaque data with either a node or an adjacency. The SPB Instance Opaque ECT algorithm sub-TLV is carried within the MT-Capability TLV (with a valid SPB Instance sub-TLV). The SPB Adjacency Opaque ECT algorithm sub-TLV may be carried within the extended reachability TLV. The information elements are

- ECT Algorithm. An ECT algorithm is advertised when the bridge supports a given ECT algorithm (identified by OUI plus index) on a given VID.
- ECT Information. ECT algorithm information of variable length.

(i) *SPB Link Metric Sub-TLV.* The SPB link metric sub-TLV occurs within the extended reachability TLV or the multitopology intermediate system TLV. These two TLVs identify neighbor(s), and the SPB Link Metric sub-TLV associated with a neighbor carries the data about the link between that neighbor and the advertising bridge that all other bridges in the domain must know in order to construct the common topology view of the domain.

The information elements are the following:

- The SPB link metric indicates the administrative cost or weight of using this link as a 24-bit unsigned number. Smaller numbers indicate lower weights and are more likely to carry traffic. Only one metric is allowed per topology instance per link.
- The number of ports is the count of (link aggregated) ports associated with this one IS-IS link.
- Port identifier is the standard IEEE port identifier used to build a spanning tree associated with this link.
- Sub-TLVs can include the SPB Adjacency Opaque ECT-ALGORITHM data sub-TLV, for the purpose of extending ECT behavior in the future.

(j) *SPBM Service Identifier and Unicast Address Sub-TLV.* The SPBM service identifier and unicast address sub-TLV is used to declare service group membership on the originating node and may also advertise an additional B-MAC unicast address present on or reachable by the node. The information elements are

- A single B-MAC address, which is a unicast address of this node. It may be either the nodal address or may it address a port or any other level of granularity relative to the node.
- The Base VID (and hence the ECT-ALGORITHM) to which the following list of service identifiers are assigned.
- A list of service identifiers: -I-SID #1 to I-SID #N are 24-bit service group membership identifiers, which are all reachable by unicast frames using the B-MAC address advertised in this TLV. Each I-SID has a transmit (T) and receive (R) bit, which indicates if the membership is as a transmitter/ receiver or both (with both bits set). In the case where the tansmit (T) and receive (R) bits are both zero, the I-SID is ignored for the purposes of multicast computation, but the unicast B-MAC address must be processed.

The SPBM service identifier sub-TLV is carried within the MT-Capability TLV and can occur multiple times in any LSP fragment.

(k) *SPBV Mac Address Sub-TLV.* This sub-TLV is not used by SPBM, only by SPBV. It contains the following information:

- SR Bits (2 bits). The SR bits are the service requirement parameter from [MMRP]. The SR bits are defined as 0—not declared, 1—forward all groups, and 2—forward all unregistered groups. These bits have two reserved bits placed in front of them.
- SPVID (12 bits). The SPVID and by association its Base VID, the ECT algorithm and SPT set, which the MAC addresses defined below will use.
- T-Bit (1 bit). This is the transmit allowed bit for the following group MAC address. Its use is directly analogous to the T-bit associated with the I-SID declared in a SPBM service identifier and unicast address sub-TLV.
- R-Bit (1-bit). This is the receive allowed bit for the following group MAC address. Its use is directly analogous to the R-bit associated with the I-SID declared in a SPBM service identifier and unicast address sub-TLV.

- MAC Address (48 bits). This is a group address and declares this bridge as part of the multicast interest for this destination MAC address. Multicast trees can be efficiently constructed for each destination (R-bit set) by populating FDB entries for the subset of the SPT that connects the bridges supporting the MAC address. This replaces the function of MMRP for SPTs within the SPBV region itself.

  The SPBV MAC Address sub-TLV is carried within the MT-Capability TLV and can occur multiple times in any LSP fragment.

I-SID Attributes

Advertised with the SPBM I-SID is a pair of attributes that indicate the desired multicast connectivity:

- Transmit and Receive—wishes to multicast to all receivers in the I-SID and receive from all other transmitters in the I-SID.
- Transmit only—wishes to multicast to all receivers in the I-SID
- Receive only—wishes to receive from all transmitters in the I-SID
- Neither—does not require multicast connectivity with other members of the I-SID, or if that capability is required, all replication occurs at the ingress I-component, using a full-mesh point-to-point distribution model, which does not require any multicast state to be installed in the SPBM core.

What these attributes achieve is to produce a remarkable shorthand for the implementation of the MEF service set:

- Transmit/receive = OFF provides E-LINE or Virtual Private Wire service.
- Transmit/receive = ON provides E-LAN or Virtual Private LAN service
- Transmit/receive = OFF/ON and ON/OFF produces E-TREE service.

E-LINE and E-LAN are straightforward. What requires some elaboration is the E-TREE service offering, which has also been called "hub-and-spoke" or "layer 2 isolation."

There are numerous examples of this in both layer 2 and layer 3. BGP route targets (RTs in [IP-VPN]) can be used in a hub-and-spoke fashion, or in layer 2. Provider bridging has a feature known as asymmetric VLAN, whereby switches operate using shared VLAN learning, and a hub sends on one VLAN and receives on the other, the spokes doing the reverse. The key concept is that in Ethernet, if "I cannot flood to you OR you cannot flood to me," we cannot see each other as we cannot receive or learn. This suggests that hub-and-spoke roles at the C-MAC layer can be achieved by simply using two I-SIDs; the role of the hub is to transmit on one and to receive on the other, and performing the role of a spoke simply inverts this assignment.

The previous examples, BGP RTs and asymmetric VID, use a "receiver pruning" paradigm, which is comparatively inefficient. An advantage of SPBM is that the multicast trees generated from a spoke only go to the set of hubs, and vice versa. This provides us with, for example, fully resilient and spared BRAS/BNG access, while not permitting customers to have any layer 2 connectivity to each other. This is shown in a very stylized form in Figure 3.9, where the solid lines indicate the downstream (hub-to-spoke) paths, and the four pecked/dotted lines the upstream paths.

The MEF model of E-TREE does provide for hub-to-hub communication in addition, as this permits layer 3 resiliency approaches such as virtual router redundancy protocol (VRRP) to be employed.

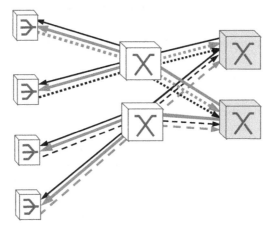

Figure 3.9   Use of multiple I-SIDs for resilient E-TREE structures.

To have SPBM to construct such an edifice, the hubs indicate both transmit and receive interest on their "hub" I-SID, and indicate receive interest only on the "spoke" I-SID. The "spokes" indicate receive only and transmit only, respectively, on the two I-SIDs in which they register interest, as in the previous example. SPBV can construct such an edifice using two Base VIDs; only hubs transmit on one Base VID and only spokes on the other.

Originally, neither transmit nor receive set attributes appeared to be useful. However, if we consider the I-SID to be the Ethernet analogy of a BGP RT, then we can identify a lightweight version of [IP-VPN], where we skip the MPLS label and put the RT in the data plane directly, again eliminating the level of indirection that label-inferred operation requires. The current specification of the I-SID does support the option of dispensing with C-MAC information and only carrying the 0x800 Ethertype and customer IP payloads, and so opening the way to this IP-VPN "tagged peer" realization, which is considered further under "IP-VPN models" (p. 167).

Aggregating I-SIDs

SPBM eliminates the need for signaling for multicast tree setup and so makes per service multicast trees tractable and resilient at a scale now limited only by the available space for forwarding state on switches. A further straightforward optimization was pointed out above; by setting the transmit/receive bits associated with an I-SID to "neither," no multicast connectivity is installed. This handles services with only two endpoints, when traffic can use the unicast B-MACs already installed or admits of using edge-based replication onto unicast connectivity if appropriate.

Despite this, it was clearly of interest to explore opportunities for further state reduction, and a number of ideas were considered to try to reduce the number of multicast trees:

- Simply using a single multicast address for each I-SID and depending on the reverse path forwarding check (RPFC) to do the requisite pruning was evaluated. It was observed first that this resulted in a less significant reduction in the amount of state in the core than might be expected. This is because in the transit layer, with a distributed set of endpoints, the number of endpoints reached through any particular port is small, often only one. The (S,G) address only has to be installed on a port for reachable endpoint source(s), and so it is only when many sources are reached through a port that (*,G) addressing confers a significant state saving. The use of (*,G) addressing also resulted in significant discard at points where the SPF trees from each edge in the I-SID overlapped.

- Another possible way forward was to find a means of aggregating (S,G) trees that shared a common set of destinations to produce a tree that could serve a number of I-SIDs. A quick analysis of the likelihood of a number of I-SID instances having an identical set of destinations for a given source was infinitesimally small for any sensible parameters and was not worth the additional complexity. A tree that was a superset and simply pruned redundant traffic was not consistent with the achievable value of the exactly pruned tree multicast at all times. Furthermore, the fit of such a tree under "adds, moves, and changes" would almost certainly degrade over time, so leading to an increasingly suboptimal topology with no obvious way of regrooming and garbage collection. Safe manipulation of large-scale routed multicast is break before make, so a hitless means of separating and reoptimizing aggregated trees simply does not exist. In summary, the likelihood of tree reuse (hence corresponding reduction of state) did not justify the additional complexity.

Although the attempt to reduce the number of multicast trees within an IS-IS area was unsuccessful, this can be done in multiarea deployments, which are discussed in a later chapter ("Multiarea," p. 171). This is because the loop avoidance strategy in these scenarios is to ensure that any service only passes between level 1 and level 2 and vice versa via a single area border bridge (ABB). Hence, it was clear that these ABB nodes at area boundaries were in effect the roots of new spanning trees, and so all the sources for a given I-SID outside an area could share a common tree inside the area. This was the one area that permitted aggregation, and it is a very important property for scaling. Multicast trees dominate the volume of forwarding state installed by SPBM, and the aggregation of multicast state at ABBs to a single tree per level 1 area per service in level 2 provides a credible scaling proposition.

SIDEBAR: *Exponential Growth of Candidate Multicast Trees as a Network Grows; Why Tree Reuse Is Implausible*

Until you do the maths, it is not intuitively obvious how quickly the number of **possible** multicast trees can grow as a network grows, and so the implications on the probability of tree reuse is also not obvious.

The equation for computing the number of combinations of exactly "k" endpoints selected from a set of "n" candidates is

$$n!/k! \times (n-k)!$$

A alternative scenario arises from enumerating all possible combinations of endpoints from a set of "n" candidates (i.e., all combinations of all $k \leq n$), which is

$$\approx 2^n \text{ for } n \gg 1 \text{ (each new endpoint } \sim \text{ doubles the possible combinations)}.$$

To produce a concrete example from this, we will pick one with some modest numbers to illustrate the point. So the number of different ways a five-site VPN could exist in a 50-node network is

$$= 50!/5! \times (45!) = \mathbf{2,118,760}.$$

The total number of possible distinct VPN topologies that could exist in a 50-node network is

$$\approx 2^{50} \approx 10^{15}$$

Both of these are surprisingly huge numbers. The implication is that with a relatively random selection of individual sites for each VPN, the number of services that even a modest network would need to support before there was a modicum of possibility of two services having identical sets of endpoints (and hence able to perfectly share multicast trees) rapidly becomes very large indeed, and hence the probability of such an occurrence in a network with a few thousand services rapidly becomes infinitesimally small.

A typical "utility" guideline for an optimization in the telecom industry is that it needs to produce a 50% improvement or reduction in cost in order to justify the complexity of implementing it and putting in place the operational procedures to exploit it. The almost nonexistent likelihood of a multicast tree being able to be shared by multiple service instances disqualifies it from consideration.

Assignment of Nodal Nicknames for Multicast Group Addresses

There are multiple ways of assigning nodal nicknames for multicast group address construction. Either administration might be used, or the task could be done algorithmically with a collision resolution mechanism. Much of the story of SPB development has been choosing the best option and closing down others, but in this case, both had desirable properties, and we felt each would play in different environments:

- The administration option obviated the disruptions inherent to a collision resolution mechanism; therefore, it would be desirable for providers.
- The algorithmic option required no configuration, and, therefore, it would be desirable to enterprises.

Therefore, the option to use either mode of operation was included in the protocol design and is described in IEEE 802.1aq.

Fast Fault Notification

Earlier, in "Elimination of Signaling from Multicast Tree Setup" (p. 76), we discussed why resiliency helpers were unable to provide significant assistance to the recovery of broken multicast trees. Therefore, SPBM is committed to the use of control plane restoration, and so we have sought to overcome the latency inherent to control plane propagation of LSPs. In a routed network today, convergence is gated by the hop-by-hop propagation of LSPs, typically imposing anywhere from a 5- to 30-ms penalty per hop even for a tuned implementation.

At first, we only considered chains of two-connected nodes, whereby any two-connected node would both sink routing advertisements and relay them directly in the data plane. This provided significant latency improvements in a scenario that we knew to be authoritatively loop free and is known to be a challenging test of control plane latency, and also carried no risk of overloading the control plane. The practical implementation was simple, any three or more connected nodes would simply not install forwarding state for the control plane multicast tree, the rest would.

One thing that became clear, during our exploration of loop avoidance and loop mitigation, was that the combination of the two would be sufficiently robust to permit the application of multicast to the control plane in a heavily meshed environment; this provided a post hoc justification for our earlier intuitive belief that minimizing periods of network instability was actually our best defense, even if at the earlier time in our deliberations the defenses had not been fully worked though.

The approach is to assign an I-SID to the control plane for the flooding of LSPs. All BEBs and Backbone Core Bridges (BCBs) in the network have both send/receive attributes on the control plane I-SID as a default, so a broadcast channel is created for the control plane in exactly the same way as for a user data service. There are obvious restrictions on the use of the broadcast channel to avoid gratuitously overloading the control plane. Only the initial announcement of a topology change is performed by the LSP initiator using the control plane I-SID. This is combined with traditional hop-by-hop flooding by the IS-IS process to guarantee reliability. As in standard IS-IS behavior, if a node receives an LSP that it has not seen before, it refloods it to all its immediate neighbors using reliable communication. So, in the extreme case where the control plane I-SID does not deliver the LSP even one hop, the traditional hop-by-hop reliable flooding will be no worse off than the current status quo.

When a node refloods an LSP received on the control plane I-SID using the reliable hop-by-hop method, it does not apply poisoned reverse to the packet. That a node successfully received an LSP on an unreliable path does not guarantee the parent successfully received and processed the packet in its control plane; hence, reliable flooding back onto the path of a potentially unreliable receipt enhances the robustness of the overall solution. The same principle applies to the children of the node on the multicast path.

When a link fails, it appears as LSP advertisements from both ends of the link. However, clearly, some portion of control plane broadcast tree will be affected. Some nodes will be in the shadow of the failure for the multicast tree rooted on the far end of the failed link. So, while a given node will be on the surviving broadcast tree from the node at the near end of the failed link, the LSP from the far end will only get partway to it on the broadcast tree and has to depend on reliable flooding to complete the journey. In this scenario, it is, however, necessary to see only one end of a link failure to generate a correct FDB for the new topology. Thus, the impact of LSP propagation delay on recovery time will typically be modest because it will affect only those nodes unreachable by multicast notifications while they are brought fully up to date by hop-by-hop reflooding.

When a node fails, the shadow of the failure is larger, and more importantly, peer nodes will need to receive a significant number of LSPs before sufficient information exists to properly compute a topology that reflects the surviving resources in the network. Even so, each computation based upon incomplete information will still be an improvement over doing nothing, and frequently the loss of a node will result in a shortest path similar to that resulting from the loss of simply the link to that node.

A related challenge to consider is this: Once a node starts receiving notifications of an event, how long should it wait before recomputing a new topology, known as the "hold-off" strategy?

The useful guideline is on receipt of the first, assume a link failure (even prior to receiving the notification from the second node connected to the link) and compute a new solution. If more than one additional LSP arrives during that interval, assume a failure of larger magnitude and delay computation accordingly in order to obtain a full set of LSP updates. This strategy is not affected by the use of fast-flooded LSPs; although more IS-IS messages will arrive, a single link failure should only cause two new LSP sequence numbers to be seen.

SIDEBAR: *The Routed Restoration Cycle*

Figure 3.10 illustrates the routed network restoration cycle.

Starting from a steady state, a fault (link or node outage) leads to a series of distinct phases before the network can return back to a steady state, each of which will have an associated time budget. From the time that the fault occurs, there will be a finite amount of time before the fault is detected by the adjacent nodes and a further finite amount of time to propagate knowledge of that failure throughout the network. A fault will manifest itself in a number of atomic transactions, the number of which is in proportion to the severity of the fault; furthermore, computation is considered to be expensive, and so a node upon receipt of a notification will typically have a hold-off time in order to collect all atomic notifications and to update the topology database before computing a new routing solution.

(Continued)

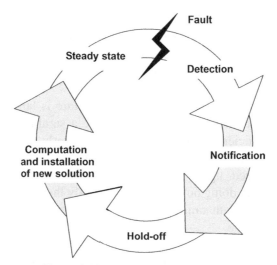

Figure 3.10   The routed restoration cycle.

Once a node has a complete database, it can compute and install a new forwarding table. Computation time is a function of the complexity of the algorithms, how frequently the node is interrupted with further messaging during the course of computation, and the raw horsepower deployed in the node. Table installation is a function of a number design and implementation factors, but in a large distributed chassis, it can become the dominant contributor to the overall time budget.

Steps can be taken to reduce time taken at each step in the cycle:

1. Delayering of the network and the use of fast-link heartbeats (IEEE 802.1ag or BFD) is pushing detection time into the milliseconds range.

2. The application of multicast to the control plane can reduce hop-by-hop control plane propagation times from roughly 5 ms per hop to frame switching time plus the speed of light.

3. The hold-off time can be tuned. A link failure will typically result in two announcements into the routing system (from the nodes at either end of the link), and a node can compute a correct forwarding table upon receipt of either one of them. A node failure will result in a larger number. Hence, a node can initiate computation upon receipt of notification of any single change and only requires hold-off once it has received three unique change advertisements.

4. The computation can be parallelized in order to take advantage of current processor technology, and messaging can be reduced to minimize the frequency of interrupting the computation.

5. The knowledge of both "come from" as well as "go to" implied by SPB computation means the amount of information required by, and communicated to subsystems in a distributed switch can be reduced with associated benefits for both transfer time and lookup table construction.

Delivered wisdom in the industry is that aspects of the restoration cycle such as notification cannot be improved upon; hence, resiliency helpers are employed to maintain degraded connectivity through periods of network instability.

However, SPB is in a position to exploit all of the techniques listed above, and as resiliency helpers (such as MPLS's FRR) raise virtually intractable problems when applied to routed multicast, the designers have focused on applying all of the above optimizations.

CONTROL PLANE: ALGORITHM ASPECTS

Consistent Tiebreaking for Loop-Free Forwarding

A key component of the routing system and the overall loop-free robustness of the network was to identify how tiebreaking should work in the presence of equal-cost paths. Much of our analysis of RPFC suggested that the consistent, symmetric tiebreaking of equal-cost paths was a key contributor to overall network robustness. The consequence is that any converged view of the edges of a cycle results in loop avoidance, as all nodes would then agree on the shortest path between any two points on the candidate cycle.

In the example presented in Figure 3.11, tiebreaking must always resolve the paths between A and C consistently, irrespective of the node doing the computation, for example, by selecting the path via B. As long as this is true, a single topology change (the shortest path to or from R changing to be via C instead of A) is safe. If it were not consistent, the change could produce a C-A path via D (while the A-C path remains via B), so forming loop CDAB.

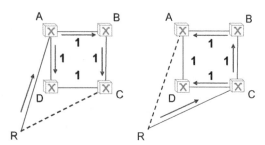

Figure 3.11   The importance of consistent tiebreaking.

The actual tie-breaking algorithm was described earlier ("Tiebreaking," p. 27). The intent here is a recap in order to explore the important properties of the algorithm. The key requirements are

- that it **never** fails to produce an unambiguous result and
- that the result is independent of the direction of computation and is independent of the position in the network of the computing node.

This can be achieved as follows:

- Compute the equal-cost paths using SPF according to Σ (link costs).
- Discard from consideration all but paths with the minimum number of hops.
- For the remainder, form a path ID, which is a lexicographically ordered concatenated list of the (unique) node IDs defining the path. Lexicographic (rank) ordering (e.g., lowest node ID first) is the crucial step in guaranteeing direction independence because it is an absolute operation on globally unique items.
- Starting from the lowest node ID, compare in turn the node IDs in equivalent positions in the path ID; the "winning" path is the one with the lowest node ID in the first position in which an exact match is not found.

This is a tie-breaking algorithm which has the interesting property that any segment of a shortest path is also the shortest path between the segment endpoints, which is referred to as "locality." That this is true can be seen by a simple symbolic example. Suppose we have an end-to-end shortest path A–H–K–T–Z, where the segment being considered is H–K–T. If there is an alternate segment H–M–T, there would be two complete end-to-end paths to consider, the second being A–H–M–T–Z. Under the tiebreaker described, using the lexicographically ordered list of node IDs, the two paths share exactly the segments A–H and T–Z, which cannot therefore play any part in tiebreaking. Therefore, the shortest end-to-end path includes the shortest segment between **H** and **T**.

The practical utility of this is computational. When computing an SPT, whenever alternate paths reconverge, the tiebreaker is used to resolve the shortest, and all state associated with the rejected path(s) may be discarded, in the knowledge that it can never be required again once having been rejected.

While the practical utility is computational, it is also a litmus test of the tractability of any additional filtering applied to tiebreaking in the development of future algorithms. Any algorithmic enhancements that do not have this property likely will have other issues rendering the algorithm unsuitable from the point of view of symmetric congruence as well. Hence, a "virtuous circle" exists, in which what Ethernet "needs" for tree construction also limits the computational requirements.

Loop Avoidance Background: Currently Deployed Techniques

Probably the most significant practical limitation of today's bridged Ethernet networks is caused by the problem of transient loops. A stable simple spanning tree has no loops. Unfortunately, distributed systems and stability do not always go hand in hand, and during instability, networks can form temporary loops. A multicast frame may be replicated many times as it goes around such a loop. This uncontrolled explosion of multicast traffic can lead, and has led, to serious network outages.

There are today several ways to deal with looping. One approach involves blocking all ports to traffic while the spanning tree is converging. Only when the tree is fully stabilized are the ports unblocked and the traffic allowed to flow. This approach, however, also requires long wait times, which can be on the order of many tens of seconds, during which traffic stops. Such long interruptions are, of course, highly undesirable since they are visible to the end users as outages, at the time of restoration as well as failure.

SIDEBAR: *STP*

Spanning tree was designed at a time when memory and compute power were scarce quantities. Hence, a spanning tree does not maintain a topology database. Its functionality is entirely transactional, whereby it simply remembers the best choice.

A simplified view of a spanning tree is that it is a "distance vector" routing mechanism. It computes a single tree with a common root. The rules are comparatively simple: The bridge with the lowest ID is the root, and the interface on which the lowest hop count to the root is received is the preferred interface. When a topology change occurs (either a link going up or down), a topology change notice is flooded to all bridges in the network. This is effectively a push of the reset button for the network. Bridges periodically exchange bridge protocol data units (BPDUs) containing the root ID and the distance to the root. When a bridge receives a BPDU with a root ID lower than the current root ID, then it changes its root ID, its distance to the root, and its preferred interface, which is the hop closest to the root.

When the bridge receives a BPDU that has a lower distance to the root than the current distance but has the same root ID, it changes the preferred next hop interface and distance to the root. When it advertises to its neighbors, the BDPU contains the current root ID, and its distance to the root + 1.

It is important to note how frugal of resources this algorithm is. A bridge only remembers three tokens of information: the root ID, the distance to the root, and the preferred next hop. The reason that any change results in a push of the reset button is because the bridge has no actual database of information; it has to reconstruct the spanning tree from scratch via the exchange of transactions with its peers. It has inadequate information available to apply any local subtlety to the required "break before make," which safe multicast needs, so break before make occurs at the network level.

Another approach to deal with the problem of looping involves adding a mechanism called a time to live (TTL) counter, identical in concept to the TTL mechanism in the IP and associated with layer 3 routing in general. TTL provides a data path field that is decremented every time a frame moves forward by one hop, essentially a countdown until a frame is discarded—somewhat like a "best before by" date. Ethernet has never had a TTL-type mechanism, and adding one through the IEEE is a nontrivial exercise that would impact deployed Ethernet hardware. Although some have proposed adding a TTL to Ethernet frames, this TTL mechanism is not an authoritative solution to a multicast loop since even a modest number of iterations around a multicast loop could potentially overload a network (to add a visual metaphor on the dangers of looping, TTL "limits the size of the crater" when the desired outcome is to avoid the explosion in the first place). Some recent papers have illustrated that such proposals as "exact TTL" matching are simply explicit versions of what the spanning tree implicitly enforces with its distance vector approach and are therefore no better than the spanning tree.

Loop Avoidance in SPB

Once RFPC was recast from being simply a mechanism to kill off persistent loops to being a key element of the multicast looping prevention mechanism, the question arose as to how robust it actually was, in particular during periods of network instability.

Investigation quickly revealed that the use of RPF was prevalent in layer 3 multicast, and many examples of vulnerabilities to looping could be identified. However, these all manifested themselves when applying the original [Metcalfe] "flood and prune" model, not his extended model where flooding is confined to the shortest path and subsequent pruning is not required. They also all involved broadcast segments because it is a comparatively trivial exercise with such segments to manipulate circumstances such that the appearance of multipath from the root is possible.

We found that if we confined SPB to run over switched Ethernet segments or point-to-point segments and stuck to the all-pairs shortest path model, whereby data path pruning was an error condition and not encountered in normal operation, then RPFC combined with link state was pretty robust.

However, an example was brought forward in the IEEE late in 2007, which does indeed create a looping scenario, but it also illustrates how contrived the formation of a loop becomes when RPFC is used as a trap for anomalies only. This example is shown in Figure 3.12, with the initial state on the left, the converged final state on the right, and the looping state in the middle. An analysis revealed the following:

1. A minimum of four nodes is needed to produce a loop.

2. The metric of one of the links in the loop exceeds the sum of the other three.

3. The need for a minimum of two nearly simultaneous topology changes: one to shift the shortest path to the root and one to close the loop.

4. Two nodes must be effectively "brain dead" (nodes B and C here). They can relay topology changes but do not instantiate the consequences of these changes in their own FDBs.

What this also revealed was that loops only become possible with the generation of scenarios where both the shortest path to the root changed, and unchanged hops occurred immediately adjacent to unsynchronized link state databases. In essence, the "recipe" is to set a cut line across the network, to freeze the network on one side in one state, then to contrive a set of circumstances such that the combination of the frozen part of the network and the part of the network that has converged on the new reality result in a loop. This pointed us in the direction of an authoritative approach, which completely eliminates any recourse to the improbability of an event that might possibly cause a loop.

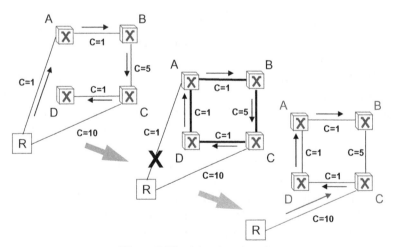

Figure 3.12 A looping example.

This ISIS-SPB loop prevention mechanism for multicast forwarding works in summary as follows. Neighbor bridges exchange digests of the topology database to check whether they have the same view of the physical topology. If they do have the same view, by inference, they also have agreement on the distance to all SPT roots. An SPT bridge only installs changes to multicast forwarding to a peer when their digests match. Traffic unaffected by a topology change, and traveling on a tree for which the distance to the root did not change, sees no interruption in forwarding. The actual condition for loop-free multicast forwarding does not require quite such a tight condition; nodes on the tree must not move closer to the root but may move further away from the root provided that no node falls below the last known (synchronized) position of its closest child.

We now amplify the logic behind this. When converged, a link state routing protocol is inherently loop free; all nodes have a complete and consistent view of the topology and construct identical trees for all roots, with a strictly monotonic increase in distance from the parent to child node as one passes down the tree.

Loops can potentially form when a topology change occurs because the topology update process of IS-IS is not synchronized; nodes act, at their own speed, as and when they receive topology updates in LSPs. Quite specifically, a loop may form if a "parent and child" relationship in the old (stable) topology becomes inverted in the new "true" topology, **before the parent and child become aware of this**. So, even though the parent is now **below** the child in the true topology, they continue to forward as before, which is now "uphill." In certain topologies, adjacent nodes that have already become aware of the true topology can begin to forward "downhill" between the old child and the old parent, thus forming a loop.

The SPB loop avoidance procedure is based on identifying these potentially loop-forming situations by examining the difference between trees formed under the old (stable) topology and trees that could be formed under the new topology, and assuming that the computing node has not synchronized with its neighbors. A node that is not synchronized with its neighbors determines, on an SPT-by-SPT basis, whether

- it is no closer to the SPT root than it was in the old (synchronized) topology;

- it has not moved further from the SPT root than any children were in the old (synchronized) topology (so it cannot have fallen below its previous children).

If both of these conditions are true, consistent application of these rules down a tree ensures that a loop cannot form, and therefore a node may continue to forward along a particular tree with no danger of forming a loop. This preserves the very desirable property of link state protocols, that traffic which is unaffected by a topology change does not need to be disrupted by that change.

If these conditions are not satisfied for a tree, forwarding on that tree must be blocked until adjacent nodes have synchronized their topology databases, and so by implication, a mutually agreed view of the "parent–child" relationships has been reestablished.

The mechanism for doing this is the exchange of agreement digests. The construction of the agreement digest is discussed in the next section; in this section, we consider how it is used to ensure loop-free forwarding.

This is now presented from the perspective of an individual node, which may only issue an agreement digest to a neighbor once it has removed the forwarding state for trees that are "unsafe" (potentially loop forming) as a consequence of the node loosing topology synchronization with that neighbor. In other words, the issue of an agreement digest is a guarantee that the node has protected itself and the network from the consequences of itself adopting a new topology view even though its neighbors may not share that view. It has already broken any possible loops.

A node must issue an agreement digest as quickly as possible to ensure that network reconvergence is as rapid as possible. By default, this requires disabling forwarding on all ports. Forwarding may, however, be enabled for a port on a specific SPT and the associated subset multicast trees when an agreement digest is sent if

Either the port is a root port (the one closest to the SPT root on this node), and

- since last synchronized on this port, an agreement digest has not been issued from this port for a topology in which
  1. this node is or was closer to the SPT root, or
  2. this port was a leaf-facing port (and hence it has changed its role)

Or the port is a leaf-facing port (known as a designated port in IEEE), and

- since last synchronized on this port, an agreement digest has not been issued from this port for a topology in which
    1. this node was further from the SPT root than was the leaf-facing port's child when last synchronized, or
    2. the port was a root port (and hence it has changed its role).

SPB ends up with unique properties with respect to failure scenarios when compared to solutions offered to date. Neither is it completely disruptive in the form of port blocking, the behavior offered by the STP hitherto used in Ethernet, nor is it dependent on demonstrably incomplete mitigations against frame replication, such as the TTL counters used in routed networks. Under failure, connectivity is maintained for unaffected paths, albeit affected paths are disrupted. However, as should become clear during the course of this book, the behavior of a connection-oriented routing system when compared with the signaled reconstruction of multicast connectivity ensures that the actual disruption even to affected paths is minimized.

In this discussion of loop avoidance, we have explicitly considered only a single forwarding plane and a single topology. In the light of the earlier discussion of multiple planes and ECMT ("The Meaning of the VLAN in PBB-TE and SPBM," p. 63), it may legitimately be asked whether a single plane treatment is adequate. The short answer is "yes," because at the frame level, each forwarding plane is associated with a single B-VID in SPBM or SPVID in SPBV. Every frame can and does transit the SPB network on such a single VID, and within the scope of this discussion, no mechanisms are available to "leak" frames between VIDs. Accordingly, if the procedures considered above are applied independently to each forwarding plane, then collectively no frame looping can occur in the network; however, many forwarding planes are implemented.

Agreement Digest Construction Details

The requirements that must be met by the topology agreement digest are:

- to summarize the key elements of the IS-IS link state database in a manner that has an infinitesimal probability that two nodes with differing databases will generate the same digest:
  - even when many elements of the database are likely to be highly correlated (e.g., the OUI values in MAC addresses, link metrics);
- to have a very low incremental computation overhead because in general, link failure and repair are isolated events, and a single event should not require complete recomputation.

To achieve this, the topology agreement digest field comprises six elements:

- the topology digest format identifier
- the topology digest format capabilities
- the topology digest convention identifier
- the topology digest convention capabilities
- the topology digest edge count
- the computed topology digest

The topology digest is carried as a structured 32-byte field in the IS-IS-SPB digest sub-TLV.

The first four fields are provided to preserve extensibility in digest selection; all but the topology digest convention identifier are set to zero in the current version of the standard.

The topology digest convention identifier indicates the strength of loop prevention being implemented in the node transmitting it as follows:

1. indicates that the transmitter will not forward until an agreement digest match occurs.

2. means the transmitter will continue loop-free forwarding of both multicast and unicast traffic up to the limits of change described earlier ("Loop Avoidance in SPB," p. 111).

3. means the transmitter will continue loop-free forwarding of multicast traffic up to the limits of change described in the previous section and will continue forwarding of unicast traffic

unconditionally, relying solely on RPFC for loop mitigation for unicast traffic.

The topology <u>digest edge count</u> is a 2-byte unsigned integer. Its purpose is to provide one component of the topology digest, which is simple to compute and powerful in detecting many simple topology mismatches. In the light of the use of a strong hash for computation of the computed topology digest, the edge count can be seen as a historical hangover from when a simpler multiplicative hash was envisaged.

This value is the sum modulo 2^{16} of all edges in the SPB region. Each point-to-point physical link is counted as two edges, corresponding to its advertisement by IS-IS in an LSP flooded from either end of the link. In the case of shared media, the number of edges will be defined by the point-to-point links joining each connected SPB bridge to the IS-IS pseudonode used to represent the shared medium within the routing system.

The overall procedure for constructing the <u>computed topology digest</u> is

- to form a signature of each edge in the topology by computing the MD5 hash (RFC 1321) of the significant parameters of the edge, as defined below and
- to compute the digest as the arithmetic sum of all edges in the topology.

Although MD5 is widely reported to be cryptographically compromised, this is not relevant in this application because there is no motivation for an attack. What is required is a function exhibiting good avalanche properties such that signatures with potentially very similar input parameters have an infinitesimal probability of collision.

This strategy also allows the computed digest to be incrementally computed when the topology changes by subtracting the signatures of vanished edges from the digest and adding the signatures of new edges. In general, the signature of an edge therefore needs only to be computed once, when it is first advertised, which satisfies the desire for very low computation overhead as a result of a single link change.

The input message to the MD5 hash for each edge is constructed by concatenating the following fields in order, with the first field being the beginning of the message:

1. The bridge identifier (bridge priority ‖ bridge sysID) of the bridge advertising the edge with the numerically larger bridge identifier value (8 bytes)
2. The bridge identifier of the bridge advertising the edge with the numerically smaller bridge identifier value (8 bytes)
3. A variable number of 8-byte 3-tuples, one 3-tuple for each MTID declared in IS-IS. The 3-tuples are declared in descending order of MTID value, with the largest MTID declared first.

Each 3-tuple is constructed by concatenating the following fields in the order below:

1. A 2-byte field containing the 12-bit IS-IS MTID value (defined in RFC 5120) in the least significant bits, with the 4 most significant bits of the field set to zero
2. The SPB link metric for the edge in this topology, which has been advertised by the bridge with the numerically larger bridge identifier value (3 bytes)
3. The SPB link metric for the edge in this topology, which has been advertised by the bridge with the numerically smaller bridge identifier value (3 bytes)

If an edge is not present in a topology, its SPB link metric is set to zero in that topology.

The value of the topology digest is the arithmetic sum of all of the signatures returned by presenting every edge message to MD5, treating each signature as an unsigned 16-byte integer and accumulating into a 20-byte integer. Every physical link is seen as two edges, one advertised in an LSP by each bridge comprising the adjacency, and formally, the topology digest includes both. Figure 3.13 summarizes the topology digest construction process.

Although the topology digest contains signatures for both edges associated with each link, the construction defined above ensures that these are always identical. An implementation may therefore choose to compute a single signature per link and then double it before accumulating it into the topology digest.

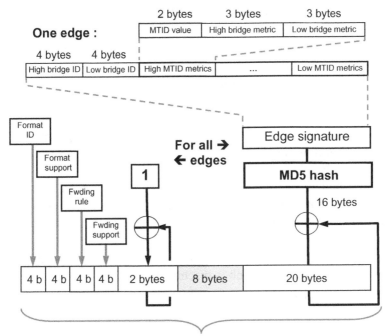

Figure 3.13 The SPB topology digest.

Load Spreading: Equal-Cost Trees

While the generation of minimum-cost SPTs has significant benefits for SPB, it is comparatively trivial to generate use cases where a single mesh solution strands significant network capacity. For example, the tiebreaking in the dual hub-and-spoke case universally means that one hub sits idle due to the definition of minimum cost used by the algorithm to construct the SPTs.

After analysis, it was realized that it was possible to produce multiple variants of the original tie-breaking algorithm (lexicographic ordering with the lowest node ID first). The simplest variant is to lexicographically order paths with the high node ID first and then to select the highest nonmatching entry. These two bookend variants preferentially selected paths that have a significant diversity across the network and so became the basis of the initial SPB load-spreading mechanism.

Initially, when considering SPBM, we had tried to avoid load spreading entirely because hop-by-hop load spreading (known as ECMP in the IP literature) is incompatible with data plane OAM and multicast congruence. Despite the fact that only one shortest path could exist in SPB to a given destination in a given VLAN (B-VID in the case of SPBM, SPVID set in the case of SPBV) for go–return congruence, we observed en passant that this did not preclude some form of edge-based load spreading. This would further retain the merit that OAM works consistently because the path per B-VID between any two edges is both unique and end to end.

The challenge to achieve "hot-spot free" load spreading in what is effectively a connection-oriented routing system has ended up with the generation of a number of approaches. Two design decisions arose:

1. how to generate multiple topologies that maximized diversity of connectivity, which is discussed now; and
2. how to assign traffic to each, which is discussed in the next section.

The choices identified for generating multiple topologies were

- using multitopology routing and assigning different metrics to links in different topologies;
- ranking equal-cost paths by leveraging some existing administered identifiers and assigning topologies according to ranking; and
- the addition of extra information to facilitate the generation of diverse paths, but ideally avoiding the computational overhead of multiple topologies.

In each case, the individual topologies would each be represented by a B-VID for SPBM.

When considering the merits of each approach, the further complication arose because we could be dealing either with comparatively arbitrary topologies, such as a carrier network, or with highly regular hierarchies, such as data center networks. The latter are so regular that one industry direction is to select control behavior optimized for very specific topologies.

One key observation emerged, and this was that the number of path permutations independent of diversity is combinatorial, but the actual number of link diverse end-to-end paths is in proportion to the maximum breath of connectivity at the widest part of the network. With SPB requiring unicast/multicast congruence and bidirectional symmetry, we were drawn to techniques that found and used link diverse paths when they existed.

When considering the approaches, multitopology routing had the advantage of providing an apparent degree of operator control, and it would provide the capability to load spread in asymmetric topologies. Unfortunately, manipulating metrics is not actually easily predictable in terms of the effect on the traffic distribution matrix because the metrics are visible globally. This would also require configuration to assign different link metrics, and we would have to incur the overhead of running the all-pairs computation twice or more.

Ultimately, several things factored into preferring that multiple equal-cost paths should be found using tiebreaking based on lexicographically path ordering despite it being an "opportunistic" technique:

- It would require no configuration.
- It only required a single all-pairs computation.
- The metro topologies we had seen from several providers were symmetrical hub-and-spoke configurations, where the ranking option would have the most benefit effectively "for free."

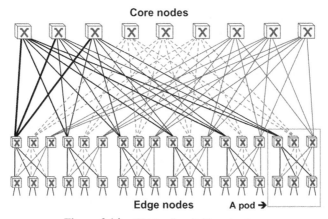

Figure 3.14   "Fat tree" switching structure.

The result was that we went initially with a philosophy of load spreading being a zero configuration tool, used simply to reduce the amount of intervention it took to properly operate the network.

The availability of a diverse path pair (low- and high-ranking bookends) seemed adequate for carrier metro deployments, with their typically limited breadth of connectivity.

However, this was not going to be a credible technique for data center applications, which typically use "fat tree" architectures, or related variants, and in which scaling is achieved by increasing the **breadth** of the network and thereby offering more and more parallel equal-cost paths between any two endpoints in a highly structured organization. Figure 3.14 shows a fat tree structure, with a very modest six ports per switch. The key point to note is that each core node offers an equal-cost route between any pair of endpoints, and the network is scaled by increasing the number of core nodes. To address this, a number of techniques to handle large numbers of equal-cost routes were explored and subsequently dismissed:

1. Relative rankings, such as second lowest, second highest had a number of problems:
   - Creating a chain of path ranking dependencies meant that a failure in a path could have an impact on otherwise unaffected paths.
   - When moving beyond trivial networks, the actual degree of diversity between, for example, the highest and second highest path was actually quite small:
     - Frequently, the majority of links in two adjacent paths in a ranking would be common, within the worst case only one node being different.
   - Where path diversity was limited, the second lowest ranked path could correspond to the highest ranked path; hence, no additional diversity was obtained.
   - The actual computational algorithms could not produce consistent results as the superset of path sets inherited by a child from multiple parents could not always align rankings beyond basic low/high.
2. Pseudorandom reallocation of node IDs prior to each path ranking step was also explored. This tended to produce results that could be predicted by the "birthday problem" equation:

The expected proportion of available diverse links used
$$= 1 - (1 - 1/n)^k,$$

where n was the number of links at a given level of hierarchy and k was the number of attempts to produce a diverse path. An example result was if there were four links at a given level of the hierarchy, and the system randomly selected four equal-cost paths, only 68% of the links would, on average, actually be used. Several of the paths selected would have links in common.

Even so, this demonstrably outperformed more trivial and more correlated transformations of the node IDs, such as XOR'ing all node IDs with a common mask value and then reranking.

This leads to the realization that link diversity and not necessarily node diversity would be the class of technique required to achieve superior diversity beyond the single low–high ranking of paths. This resulted in the exploration of numerous approaches for constructing path IDs from link IDs, with the link IDs constructed from concatenating node IDs and then reranking with transformations of one end of the link ID while holding the other constant. This did exhibit some linkage between iterations of path ranking such that there was a higher probability of diversity with a previous ranking. This technique at the time appeared to be the high watermark of what could be achieved in a single pass of the database and masking of path IDs.

Having established the limits of what seemed to be possible with automatic techniques for identifying multiple paths, it is clear that the highly structured nature of fat tree and similar topologies offered a means to significantly increase the path diversity with a minimal amount of administration. In such architectures, the selection of just a core node fully defines the path between any two endpoints. So, when revisiting the technique of transforming the node IDs by XOR'ing all node IDs with a common mask value and then reranking; as was mentioned above, this does not perform very well when used with pseudorandom variables. However, if the common mask is selected to be the node ID of each core node in turn, each core node is selected as the low path tiebreaker precisely once, irrespective of the value of any other node IDs.

It was becoming very apparent that the original low-rank, high-rank tie-breaking algorithm, originally perceived as "the answer," was actually only the first of a suite of possible algorithms, some especially well matched to particular classes of topology. In order not to delay the standardization of shortest path bridging, or to restrict subsequent developments, an extensible framework was proposed and adopted, which was populated with some known algorithms but allowed the incremental addition of later ones.

Load Spreading: Assignment of Load to Trees

The other decision needed was how to assign load to each topology created. Here, we went with the simplest solutions, which were, for SPBM,

1. spreading on the basis of community of interest with configured assignment of an I-SID to a specific B-VID,
2. algorithmic assignment of I-SIDs to B-VIDs, and
3. per-flow load spreading.

Assignment of I-SIDs to specific B-VIDs (option 1) was consistent with the existing PBB approach (intended for managed operation by a network operator). It also readily permitted services to be instrumented since I-SIDs could inherit OAM properties from the B-VID. Finally, it minimized state in the B-MAC layer since the set of (S,G) trees for a given I-SID did not have to be replicated in every topology; a multicast tree for a BEB/I-SID tuple only appears in the B-VID to which the service is assigned, and so load spreading did not involve fully duplicating state in every topology.

Algorithmic assignment of I-SIDs to B-VIDs (option 2) can be achieved by assigning, for example, the even I-SIDs to one B-VID, and the odd ones to the other:

- This also added no overhead to the routing system, and
- it ensured that I-SID to B-VID mappings were synchronized network-wide.

The latter is especially important when one considers that unsynchronized bindings could result in the congruency property being disrupted. If one I-SID endpoint was not in the same B-VID as

another, connectivity between the two could not be properly constructed, and even if tables were correctly populated, RPFC would
obstruct operation as the connectivity in each direction would be in
different B-VIDs. Finally, unsynchronized bindings are ultimately
operationally confining as they eliminate the possibility of seamless
service migration.

What is not standard but would also be possible is load spreading
of flows (option 3). This addresses enterprise applications, when the
number of IP subnets may be modest, and so per I-SID spreading could
be too coarse a granularity. Further, the multiple paths delivered by
ECMT exhibits many of the properties of a virtual multipoint link
aggregation (LAG) (see p. 121), so that per-flow load spreading at edge
BEBs is a practical option, with each flow being sent over one and only
one of the available multiple paths. The implementation penalty is that
the I-SID trees need to appear in all B-VIDs, but this is not considered
a scaling impediment in enterprise applications with the limited number
of I-SIDs which this technique is designed to address. This is similar
in concept to but simpler than "the dual-homed UNI using LAG emulation" described later (p. 121), except that the customer edge (CE) (C-
MAC) bridge of that example is here integrated into a single BEB as
the VSI associated with the I-SID.

In the case of SPBV, there are two distinctly different deployment
scenarios. At a modest scale, each service can be assigned to a separate
Base VID, in the same way as an IEEE 802.1ad service is assigned to
an S-VID. The achievable scale is very constrained; recall that SPVID
consumption is the product of the number of bridges and the number
of services.

In an enterprise environment, where isolation between VLANs may
not be necessary, the virtual multipoint LAG concept mentioned above
may be employed. Here, two or more Base VIDs are considered as a
LAG group, and edge-based hashing is used to select the Base VID on
which a particular flow is sent. At the egress of the SPBV region, the
SPVIDs of all Base VIDs are merged back onto a single external
VLAN. It is necessary to ensure the hashing algorithm used is symmetrical under destination and source address transposition for unicast
frames. In other words, $\{DA = X, SA = Y\}$ must resolve to the same
Base VID as $\{DA = Y, SA = X\}$ for all X and Y; otherwise, the two
directions of a flow may not follow the same path, and reverse path
learning breaks.

THE CO NATURE OF ETHERNET AND ITS IMPLICATIONS FOR ROUTING

All-Pairs Computation

Once the merits of the all-pairs shortest path computation are established by its ability to eliminate signaling from multicast tree installation, it may be an asset exploitable for other purposes; what else might it enable?

The first consequence of this thinking was that a distinct by-product of the computation was that all "loop-free alternate paths" were known, those paths passing through immediately adjacent nodes that were closer to the destination but not on the shortest path. Maybe these might be exploitable for rapid fault recovery (a sort of "FRR for SPB"). However, there are a number of complications that have precluded the exploitation of this property. The first is that it would need to assume that every node had a full forwarding table on every interface. This is true for nodal unicast MACs for SPBM in nodes that implement a common FDB, but it is not true for SPBM multicast addresses nor for SPBV SPVIDs that are installed on a "connection oriented" basis, as neither should forward frames received on any interface except that on the path to the root of the tree. The second complication is that a loop-free alternate path inevitably got us into multipath merge scenarios, which would invalidate RPFC. We could safely turn RPFC off for unicast (and it is easy to do this on a per frame basis as the DA has an explicit unicast/multicast bit), but the conclusion at the present time is that any protection solution needs to address unicast **and** multicast; hence, loop-free alternate paths had only limited utility.

So the question became "What other aspects of SPB could the all-pairs computation address?" The conclusion was both scaling and restoration performance. SPBM scaling is improved because the per-port forwarding tables can be personalized; with the all-pairs shortest path, the ingress port used by any **source** is known, and so a destination B-MAC associated with that source (a multicast address or a per-port MAC on the same service) need only be installed on that port. Furthermore, this also directly improves restoration performance because personalization will reduce the size of individual batch downloads of tables by typically an order of magnitude compared to tables, which are common to all ports on a node. This is significant because table download time dominates the actual compute time in many switch implementations.

Since I-SID information exists in the routing system for multicast computation, it can be determined whether a given pair of nodes in the network actually has any interest in each other at all. The most extreme form of this state compression is to observe that if two nodes do not have I-SIDs in common, devices on the shortest path between them do not strictly need to populate the forwarding state at all (either unicast or multicast). This was ultimately dialed back to always populating the unicast loopback addresses of all BEBs and BCBs, such that basic unicast connectivity exists for OAM, and hence for commissioning, testing, and so on; this allows, for example, link trace messages to be used without the responses being pruned by RPFC.

Partitioning and Coalescing

Although the all-pairs shortest path is computationally tractable at a significant scale using modern processors, it is not trivial, and it has always raised questions in the minds of those who have not encountered it before. Further thought triggered by this showed that it was not necessary to compute all pairwise paths through the network, and it was possible to condense the amount of computation required. The technique became known as "some pairs shortest path," as the fundamental procedure was to partition the network into sets of nodes that obviously had to transit the computing node to reach nodes in other sets but did not have to transit the computing node to reach nodes within the same set. Figure 3.15 shows how a network can be divided from the perspective of node 2 into four partitions, reached through node 2's neighbors, 3, 4, 5, and 6.

This ability to partition the network was predicated on the downstream congruency property of the tie-breaking algorithm; any portion of the shortest path is also a shortest path. This property permitted us to prove that some paths did not traverse the computing note without actually having to do the detailed computation.

The initial step was to compute the tree rooted at the computing node such that the set of nodes reachable from each interface could be determined. This partitions the network into sets of nodes (which may be null) reachable via each immediate neighbor. The next step was to determine when the shortest path between **immediate neighbors** did

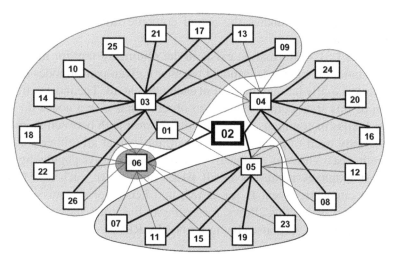

Figure 3.15   Network partitioning from the perspective of node 2.

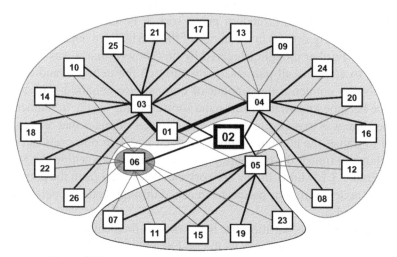

Figure 3.16   Network coalescence from the perspective of node 2.

not transit the computing node. When this is so, the full set of paths between all children of such a pair of immediate neighbors cannot transit the computing node either, so "all those pairs, shortest path" does not need to be found by the computing node. So, the sets of nodes reachable by each could then be merged.

In Figure 3.16, node 2 computes that the shortest path between two of its neighbors, nodes 3 and 4, is via node 1, and therefore node 2 knows that no shortest path between the upper gray regions can transit it, and the regions can therefore be grouped from the point of view of transit routing.

When all possible "coalescing" of sets of nodes that cannot transit the computing node has been performed, the all-pairs SPF computation then needs to be performed for the nodes in all sets but the largest set. The all-pairs computation on all the smaller sets automatically finds transit paths between them and nodes in the largest set, and the reverse path symmetry property means that the return path for these is known without further computation. What has been completely eliminated is the tree computations for all nodes in the largest set as the root.

The absolute extreme cases can both be envisaged:

- when no paths transit the computing node; this can be determined by a small number of Dijkstra computations, after which the computation is complete. This circumstance is usually found at the edges of a network, where the cost and hence the computation performance of the large number of nodes is most likely to be constrained, and so the "benefit multiplier" is the greatest;

- when all paths transit the computing node (which is therefore a single node at the center of a star topology).

Optimizing Point-to-Point Scenarios

Another and so far unexploited realization was that, with the correct AAA infrastructure, SPBM is actually a near-real-time fulfillment mechanism for nomadic access to a "home" LAN segment: "remote access at and over layer 2." One implication of this was that we would expect a significant number of E-LAN services, which only served two SPBM UNIs, and so installation of multicast state was not required, and all that was required was knowledge of the nodal B-MACs associated with the I-SID. This led to a small modification to the algorithms to cause all shortest path bridges to simply use unicast connectivity when an I-SID only appeared twice in a network. This also vindicated the approach of fully populating the tables with the nodal loopback MACs, as no forwarding table churn would be associated with commissioning or decommissioning a two-site E-LAN service.

CHAPTER 4

Practical Deployment Considerations

IN-SERVICE UPGRADE AND SERVICE MIGRATION

There are numerous reasons for making in-service changes to network configuration; one important reason is to modify how the traffic matrix is distributed within the network in order to relieve "hot spots." One way to do this is to migrate customer VLANs between ECT sets. Another is the introduction of a new ECT algorithm, to make a new set of paths available.

SPBV implements a VLAN directly associated with a Base VID using an SPVID set. As such, the only practical migration scenario would be to change the ECT algorithm associated with a Base VID. This is nontrivial and provides a transient load on a scarce resource, the virtual LAN identifier (VID) space.

For SPBV, the assignment of one of the bridges to the spanning tree algorithm forces all bridges in the SPT region to use the spanning tree algorithm. This allows the other bridges in the region to be modified for the new algorithm, which will not take effect until the bridge set to the spanning tree is then transitioned to the new algorithm. The technique is "hitful" in that service is disrupted twice for the duration of network convergence.

SPBM has separated the service from the ECT algorithm with a level of indirection (the mapping of the I-SID to a B-VID/ECT algorithm), where each B-VID defines a fully meshed forwarding plane.

802.1aq Shortest Path Bridging Design and Evolution: The Architect's Perspective, First Edition. David Allan and Nigel Bragg.
© 2012 the Institute of Electrical and Electronics Engineers. Published 2012 by John Wiley & Sons, Inc.

This allows a variety of in-service reconfiguration scenarios to be realized that are both nondisruptive and more frugal of state during the transition period. Examples of scenarios where this might be required are the following:

- When multiple tie-breaking algorithms are in use, each generating a different equal-cost tree set, each is assigned to a different B-VID, and I-SIDs are in turn assigned to these B-VIDs. If traffic volumes on different I-SIDs are very unequal, the technique available to rebalance the network is to reassign some I-SIDs between B-VIDs to move traffic off heavily loaded links.
- In "fat tree"-like architectures, scaling a network can involve increasing the number of core bridges and, hence, the breadth of connectivity through the network. This requires the introduction of new algorithm variants and associated B-VIDs to exploit the new connectivity.
- In other scenarios, it may be desired to introduce new tie-breaking algorithms, better tuned to the evolving use of the network.
- Finally, it may be desired to migrate a network between forwarding modes. For example, it would be sensible to upgrade a network originally installed to use PBB (with spanning tree) to SPBM. Another scenario would be the in-service upgrade of a point-to-point connection, delivered using an engineered PBB-TE trunk to add further point(s) of connection, which would then be delivered using an SPBM LAN segment.

In all cases, the solution is similar. It exploits the fact that SPBM has a clean separation between the endpoint address (the B-MAC) and the path identifier (the B-VID). In each of the cases above, the same basic procedure is followed:

- The new forwarding planes are configured on the live system but with no traffic assigned to them (i.e., no I-SIDs are associated with the new B-VIDs). Once the incremental configuration has been installed and is synchronized in IS-IS, a prudent operator is likely to verify the performance of the new forwarding planes before considering them ready for service.
- At this point, I-SIDs can be transferred from their previous B-VID(s) to the new one incrementally, either one by one or in

batches. First, the set of I-SIDs is configured to be associated also with a new B-VID, in receive-only mode; at this stage, each endpoint will receive from either the old or the new B-VID, although nothing has yet changed in the forwarding plane. When a node finds itself on the shortest path between any two nodes with interest in the I-SID, it will never at this stage see any with the transmit bit associated with the new B-VID set; hence, it will not put any forwarding entries in the filtering database (FDB).

- Then, cut-over can be performed on a node-by-node basis for each I-SID (set). The I-SID transmit attribute is set in the context of the new B-VID, which installs the required new multicast forwarding state. The I-component is then instructed by management action to swap transmission to the new B-VID. In the forwarding path, this action can be and usually is atomic, and so the changeover is lossless, and will be hitless unless there is significant difference in latency between the B-VID paths.

- Once the nodal changeover is complete, the I-SID(s) are made receive-only on the old B-VID to recover the now-unused forwarding path resources.

- This procedure is repeated at each node in turn, in any order desired. There is no requirement for any defined synchronization between nodal cut-over because all nodes have been configured to "send on one B-VID, receive on either." The only SPBM attribute lost during the transition is congruence, and so it is undesirable to prolong the transition period unnecessarily. Once the cut-over has been completed on the entire network, the entries of the transitioned I-SIDs associated with the old B-VID can be deleted.

In summary, the SPBM model, with multiple B-VIDs each defining a single fully connected forwarding plane, allows lossless and usually hitless network regrooming using a single common set of operational processes for a wide variety of scenarios.

DUAL HOMING

For conventional Ethernet, dual homing is something of an oxymoron because when deployed, the Spanning Tree Protocol (STP) can never simultaneously unblock both links to create two paths between the source and the destination as to do so would immediately form a loop. Nonetheless, it is frequently desired to install redundant capacity so that a link failure cannot sever a network. The important question then becomes whether the use of link state routing can make better use of such capacity under fault-free conditions than does spanning tree.

We can identify several distinct scenarios:

- The most straightforward is a dual-homed edge node running SPBM (a Backbone Edge Bridge [BEB]) and peering with SPBM Backbone Core Bridges (BCBs). In this case, which is an **NNI**, the natural operation of link state routing builds trees, which will use whichever link offers the shortest path for a given destination.

- The second case is represented by a dual-homed NNI with a node that is not running SPB (which might be a conventional Ethernet bridge). This is a degenerate example of a more general case, that of multiple connections between the SPBM domain and a shared LAN segment. A later section, "Shared Segments" (p. 141), discusses the way this scenario may be addressed, and so this will not be considered further at this point.

- The challenging dual-homing scenario is when an edge node has a dual-homed **UNI** onto an SPBM BEB because SPBM then has no knowledge within the routing system of the customer media access control (MAC) addresses, which are those seen on the UNI. We consider this case first.

- Conventional Ethernet does have link aggregation (LAG) as a technique for bundling multiple links into a single point-to-point connection. This is used for load spreading but also provides redundancy. The second technique we consider provides resiliency by emulating a LAG UNI when viewed from the edge node, but having SPBM offer link and nodal resilience when such a LAG is connected to an SPBM network.

- There is a recent ITU-T standard [G.8032] specifying the operation of resilient Ethernet rings without the use of spanning tree. The use of this technology would allow deployment of a more sophisticated form of the dual-homed UNI discussed in the next section, in which the access connectivity was self-repairing, without any visibility of a failure in the SPBM domain, and only the failure of the active BEB would require the point of attachment of the service to be moved.

As a general principle, it is very desirable that faults outside a network region (such as a fault on a dual-homed customer interface) should not be visible inside the region, to avoid propagation of disruption and control plane load. However, achieving this invariably requires network resources at the region boundary dedicated to achieving this isolation. In some scenarios, limited disruption may be preferred to dedication of resources. We therefore now discuss two dual-homing techniques, the first of which is a "minimum resource" approach, and the second offers "minimum visibility."

The Dual-Homed UNI

In the case of a dual-homed UNI, the choice of control protocol available in practice for loop prevention is STP, which addresses a general mesh, or G.8032, which is focused on ring topologies. In the Carrier Ethernet service environment, participation in many STP instances (i.e., one per customer) is undesirable, and it would be highly preferable to have an SPBM-only solution for the dual-homed UNI. For the same scaling reasons that participation in many STP instances is deprecated, it is also undesirable to make the customer edge (CE) node visible to SPBM directly (the required IS-IS construct would be a pseudonode per CE, which would be a significant scaling issue). A way in which this control plane load can be avoided is illustrated in the example scenario shown in Figure 4.1.

Node C is a CE switch dual-homed to SPBM nodes A and B and presents a tagged UNI with two VLANs to these nodes. Each VLAN can be treated as a separate service from the perspective of the SPBM core. Nodes A and B each have two UNI endpoints—one for each service—and detect failures of their respective links through physical layer mechanisms (LOS, autonegotiation RFI bit, etc.). Nodes A and B have no control plane interaction other than through SPBM, and they each see a connection to the pair of broadcast LAN segments defined by VLAN x and VLAN y.

In normal circumstances, both C-A and C-B links are active, and we want traffic from C belonging to one of the two VLAN services to enter the SPBM core via node A and traffic on the other VLAN service to enter via node B. We assume here that for VLAN X, node A is primary, and for VLAN Y, node B is primary. Because both links are active and no control plane protocol is blocking C-A or C-B, multicast traffic from C on both the VLANs will be received by both SPBM nodes. The SPBM UNI, however, drops traffic on a dual-homed endpoint by blocking the port on the service VID if this node is "secondary" for this service and it has information (through IS-IS) that its primary partner's UNI connectivity is fault-free. The SPBM advertisements from nodes A and B under fault-free circumstances are shown in Figure 4.2. Notice that the B-MACs for each I-SID are the **port** addresses of the UNI, so that they can be withdrawn individually and that they are the same on both nodes; the "owner: true/false" flag determines which is used for traffic. To avoid any possibility of frame duplication or looping during recovery from failure, "owner: false" has two states: active and inactive.

If the C-A link fails, node A will detect the failure and withdraw its advertisement for the affected I-SIDs as shown in Figure 4.2. The solution expects that node C will also detect the link failure by physical layer mechanisms, purge its FDB of affected MAC entries, and flood

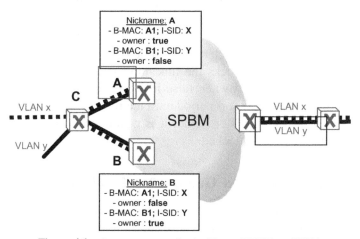

Figure 4.1   Announcement of a dual-homed UNI into SPBM.

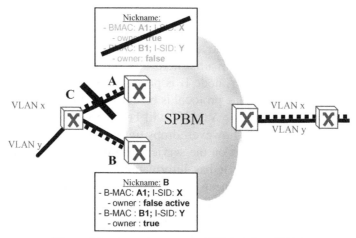

Figure 4.2 Announcement of a UNI fault into SPBM.

such MACs as unknown over a link set that includes C-B to learn the new route.

In this state, VLAN X becomes unblocked on node B, and traffic received on it will start being transmitted into the SPBM core. Node B now advertises I-SID X with attribute "owner:false active." Because the port MACs of both UNIs to C are the same, remote bindings of C-MACs located "behind" node C to the B-MAC remain valid and do not have to be relearned; only the SPBM route has to be reinstalled.

Upon restoration of the C-A link, the port MAC for the UNI is readvertised by node A, which triggers node B to revert to its secondary role. However, node A blocks the link until it learns from IS-IS that node B has reverted to "owner:false inactive," signifying that it has blocked link C-B. For a short period until the routing system stabilizes, traffic may reach either B or A, but it is guaranteed that only one of the two links is ever unblocked.

The Dual-Homed UNI Using LAG Emulation

LAG has been a part of Ethernet for many years, initially standardized as a physical layer function, now moved as an unaltered functionality into the link layer. It defines the ability to treat multiple point-to-point links as a unit, to accommodate traffic growth, and to provide resiliency. To do this, a set of links is defined as a LAG group, and the forwarding processes at each end spread load over the links which are functioning at any time, by hashing destination and source MAC addresses to select the link to be used for each frame. Solutions have emerged in the industry that use LAG administrative capabilities to force a form of protection switching, but thereby stranding access capacity similar to the behavior seen with STP. More recently, this has been revisited in order to get to the ability to once again use multiple uplinks in a distributed LAG.

A LAG with multiple active links has a key characteristic needed for dual homing; each frame is sent precisely once between the two endpoints, frequently on the basis of hashing frame information to ensure "flows" (all transactions between a pair of nodes) receive common treatment.

However, in the dual-homed case, one "endpoint" must actually consist of a pair of BEBs, and this pair must emulate LAG behavior when viewed from the client endpoint. How may this be done with SPBM?

The fundamental principle is captured in Figure 4.3. The LAG function is implemented by SPBM as a pair of **multipoint LAGs** (also called sLAG) connecting all service endpoints, which can also be viewed as a pair of virtual LAN segments (shown solid and dotted). This preserves the property that each client frame is delivered once and once only. However, to make this work properly and resiliently, there are certain "devils in the detail," which are now described.

There is a need to synchronize pairs of BEBs for a number of reasons, of which the most fundamental is C-MAC learning. Recall that in SPBM, a BEB associates a C-MAC with the B-MAC of the remote BEB which sent the frame, which is the virtual equivalent of conventional Ethernet's learning to associate a source MAC with its port of arrival. In a LAG environment, it would be unwise and burdensome to rely on synchronization of the spreading function on CEs; indeed, it could be impossible if, for example, one of the CEs had a true LAG with a host BEB as one of its dual-homed access points. Accordingly, frames can travel in one direction on the solid multipoint LAG in Figure 4.3, and the return path for the same C-MAC pair could follow the dotted multipoint LAG. So, unless BEB pairs (e.g., BEB-1, BEB-2) share knowledge of MAC bindings, there are pathological scenarios where customer frames could be "flooded as unknown" indefinitely.

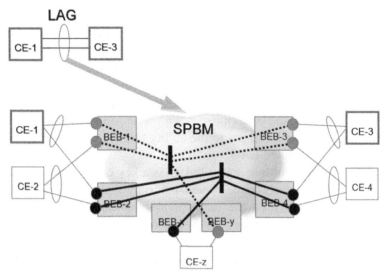

Figure 4.3 SPBM delivering "multipoint link aggregation."

BEB pairs, therefore, are required to set up a private trunk connection over SPBM and to use this for mirroring their C-MAC FDBs.

The other function of this trunk is to share knowledge of the state of LAG links to CEs, so LAG behavior can be correctly emulated under fault conditions. As we shall see shortly, the status of the CE ⇔ BEB links is hidden from the routing system and is never advertised. This is a deliberate decision, consistent with principles outlined above, and taken to maximize the scaling capabilities of SPBM, as a substantial number of CEs will typically be hosted on a pair of BEBs.

Figure 4.4, showing now a single endpoint of a multipoint LAG pair in the same graphical convention as above, illustrates how this can be achieved. The text boxes show the key information flooded in IS-IS and associated with the set of I-SIDs being handled by this BEB pair. BEB-1 advertises the I-SIDs in the context of a virtual B-MAC that uniquely identifies the LAG—B-MAC_sLAG-1 is associated with the solid B-VID-1—so that unicast and multicast traffic, which passes from the CEs via BEB-1, always gets sent on B-VID-1, and returned traffic on B-VID-1 is always routed to BEB-1.

BEB-2 performs an identical role using the dotted B-VID-2 and connecting to the other LAG'ged access link to the set of CEs. Notice

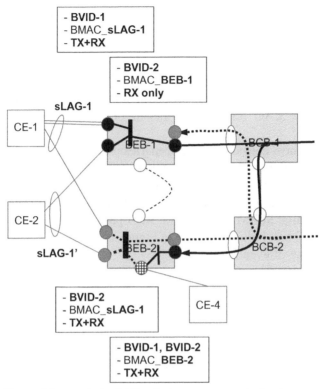

Figure 4.4 Multipoint LAG endpoints: structure and advertisements.

that the B-MAC_sLAG-1 advertised is the **same** for both BEBs (but over different B-VIDs, so no ambiguity arises in IS-IS). This address, in essence, proxies for the set of CEs hosted by the BEB pair and means that C-MACs can be learned against a common B-MAC, irrespective of the route (i.e., B-VID) that the LAG function determines they happen to use.

The next nuance is to advertise the "secondary" B-VID for a BEB (so B-VID-2, dotted, on BEB-1, B-VID-1, solid, on BEB-2) as "receive-only" for the I-SIDs associated with the multipoint LAG. This is prepositioning state in the associated multicast trees for rapid restoration, and we discuss this shortly. The state and traffic burden is modest; because the I-SIDs are advertised "receive-only," so no new multicast trees are built, and each instance adds a single leaf to the existing trees. Under fault-free conditions, traffic arriving on these secondary B-VIDs is dropped at the NNI of the BEB.

In this scheme so far, we have focused exclusively on CEs using LAG. However, BEB-2 has a single-homed CE (-4). The I-SIDs associated with this are advertised in the context of **both** B-VIDs and a **nodal** B-MAC (single homing does not require a proxy address). BEB-2 now performs the LAG functions on behalf of CE-4 as part of its UNI processing.

We can now consider failure scenarios. Faults in the SPBM domain are handled entirely within that domain; if SPBM connectivity to both BEBs of a pair exists, IS-IS will find and use it. As mentioned earlier, the converse is also true, and to avoid exposing per UNI state to what may be a large SPBM network, the BEB pair executes local repair. An access LAG link failure is shown in Figure 4.5. This causes the following sequence of events:

- Using physical layer mechanisms, BEB-1 detects the link failure.
- BEB-1 alerts BEB-2 to its loss of connectivity to CE-1 using the private synchronization trunk.
- Unicast traffic for CE-1 arriving at BEB-1 on the B-VID represented by the solid line is tunneled to BEB-2 over a point-to-point SPBM connection reserved for this purpose.
- BEB-2 removes the B-MAC encapsulation and relays the traffic onto its connection with CE-1.
- At the same time, BEB-2 unblocks the preinstalled receive-only termination for multicast traffic on the solid B-VID-1, for traffic destined to CE-1 only, so that multicast frames that would normally be relayed via BEB-1 are now relayed to CE-1 via BEB-2.

By this combination of redirection of unicast traffic, interception of multicast traffic, and B-MAC stability under failure which obviates the need to flush learned C-MACs, the entire SPBM domain can be unaware of local failures on access links.

The final class of failure is complete BEB failure. In this, BEB-2 fails, and BEB-1 detects both failure of its synchronization trunk and loss of IS-IS visibility of BEB-2. At this point, BEB-1 assumes the identity of BEB-2 and advertises the set of I-SIDs for their LAGs as

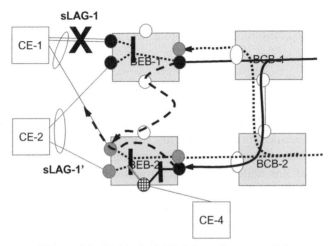

Figure 4.5 Multipoint LAG: failure of an access link.

transmit and receive on both B-VIDs. If the BEB itself has not failed, and instead a catastrophic loss of SPBM connectivity has completely disconnected it, both BEBs will each assume the other's identity. What residual connectivity remains is uncertain because the SPBM domain is severed:

- But because it is severed, we can say that whatever frames are delivered will be delivered precisely once and that loops will not form.

SHARED SEGMENTS

So far, the exposition has assumed point-to-point connectivity between bridges in the SPB domain and ignored the traditional multiple access shared segment. This chapter considers this topic since, although it is reasonable to ignore a physical multiple access LAN segment when nearly all physical connectivity is fiber, an SPB overlay of an **emulated** LAN segment (as is offered by, e.g., [VPLS]) is a real deployment scenario. We describe the solution to this challenge, but it is "not a done deal" in standards because it needs modest extensions to the Ethernet forwarding path for SPBM. We first explain how IS-IS itself already supports LAN segments and then enumerate the potential looping challenges in the forwarding plane which LAN segments present. We conclude by outlining the extensions to Ethernet needed to mitigate these issues.

IS-IS Support of Shared LAN Segments

Shared and switched LAN segments provide an efficient form of any-to-any communication for scenarios where metrics are not required. However, this communication model presents an issue, that of dealing with a set of link state nodes which traditionally need to form point-to-point adjacencies with each other in order to share their link state information. Since each node on a LAN segment can speak to each other directly, there becomes a strong motivation to create adjacencies between the IS-IS speakers in a way that avoids the formation of $O(N^2)$ point-to-point adjacencies between each IS-IS node connected to the same LAN segment.

In order to address this issue, IS-IS was designed to model LAN segments as star topologies, both from the perspective of control and metrics, and each node creates only one adjacency to a virtual IS-IS speaker called a pseudonode, as illustrated in Figure 4.6. This model greatly enhances the control plane scaling since creating a full mesh of adjacencies between all IS-IS nodes on the same LAN segment would be too inefficient. Therefore, the function of a pseudonode is to synthesize a description of each node attached to the same LAN segment as if it had a single adjacency to the pseudonode.

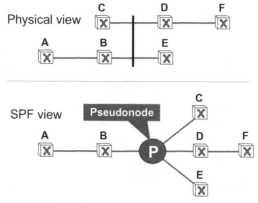

Figure 4.6   IS-IS use of the pseudonode to model broadcast segments.

To facilitate this logical topology, one node on a LAN segment will be elected as the designated intermediate system (DIS) based on a metric such as DIS priority, or by the highest value of MAC address attached to the LAN segment if the priority is the same. The DIS node is responsible for creating and maintaining the pseudonode by generating link state advertisements for the pseudonode, separate from its own, that list all other nodes on the broadcast network as being adjacent to itself in its adjacency list. Each node on the LAN segment will also announce that it is directly connected to the pseudonode. To ensure that link metrics are used properly, each node announces its own LAN interface metric as the cost to reach the pseudonode, while the pseudonode announces the cost from itself to each real node as a zero-cost link.

The DIS responsibility is to announce the pseudonode LSPs identified as distinct from its own. It does this by creating a second set of LSPs that have a System ID of the DIS plus a pseudonode ID to show that the LSPs are originating separately from the announcing IS-IS node. Each node announces that its neighbor is the pseudonode, not the DIS, where the neighbor's address is the System ID of the pseudonode.

The pseudonode acquires an extra responsibility in SPB, that of being the anchor point for topology synchronization for loop avoidance. All nodes attached to a shared LAN segment must have the same topology view before they can consider themselves synchronized and reinstall what may have been an "unsafe" multicast state. This is achieved as follows; each node on the LAN sends a topology digest to its pseudonode adjacency when it has removed any unsafe multicast state, in the normal manner for SPB. However, the pseudonode itself only advertises a new topology digest to its adjacencies when it has received an identical digest to that it has computed itself from all its adjacencies. Thus, all nodes see their LAN-facing ports as unsynchronized until all agree on a common digest value.

The use of a pseudonode does introduce a failure sensitivity, that of the DIS itself, or its attachment port to the LAN segment. IS-IS handles this in a simple manner, although the simplicity does compromise the recovery time. The DIS is elected first on DIS priority and, if that results in a tie, then on the basis of its MAC address. When nodes attached to the LAN segment detect that the DIS is no longer present, they simply rerun the DIS election procedure. The new DIS, if it has LSPs for the previous pseudonode, purges them by retransmission with a lifetime = 0 and then establishes adjacencies on behalf of the pseudonode with the other nodes on the LAN segment. Consequently, the rest of the network sees the disappearance of both the DIS and the pseudonode, followed by the appearance of a new pseudonode with connectivity to the remaining nodes on the LAN segment.

The use of shared LAN segments provides powerful mechanisms for providing interoperability and/or interworking with other Ethernet technologies. For example, a VPLS VPN can be used as a LAN segment to connect several SPBM nodes together to create a much larger network topology to meet customer needs, which VPLS could not scale to deliver on its own. Alternatively, for access to an SPB network, a shared LAN segment can be used to provide multiple points of entry, and the pseudonode concept is extended to provide a single logical representation of the access shared LAN segment to the rest of the SPB network.

Using SPBM over Shared Segments

SPBM operation in conjunction with broadcast and switched multipoint segments has been an intermittent topic of research since the genesis of PLSB. The desired goal has been that there should be no architectural restrictions on using SPBM with any Ethernet construct, despite the fact that the current provider state of the art almost universally uses point-to-point fiber interconnects. This goal has proven elusive in the general case, but much learning has ensued.

The chief learning is that LAN segments are a major and unavoidable source of transient loops, in particular when they are deployed back to back.

In Figure 4.7, any combination of events that causes any of A, B, or C to believe they are transit nodes for the tree rooted on R and then to change direction without synchronization can produce a loop.

Making SPBM work in existing VPLS deployments has been a sufficiently important topic that we have discovered that operating SPBM over a sparse and geographically separated set of switched segments has proved to be a practical and achievable goal. A particular objective has been to achieve multihomed interconnect (for resilience) without the need to block all but one link on the boundary. The basic

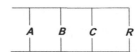

Figure 4.7   The ease of loop formation on LAN segments.

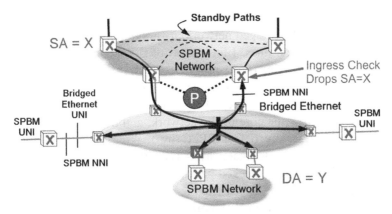

Figure 4.8 SPBM overlay of bridged Ethernet emulation.

deployment aspiration is illustrated in Figure 4.8, in which SPBM overlays a core network that offers emulated Ethernet connectivity for the B-MACs of SPBM.

A broadcast segment will naturally produce many copies of a given frame, since one will appear on every egress interface of the segment, which is why shared segments have not yet been addressed for SPBM in 802.1aq. For SPBV, this is not an issue because the SPVID of the frame naturally constrains the path the frame can take. But for SPBM, there are many opportunities for duplicate frames to "leak" into the larger network beyond the shared segment. Most duplication in this scenario is suppressed naturally by SPBM's reverse path forwarding check (RPFC). This is despite the observation that B-MACs sourced from within SPBM but unknown to a LAN segment will be reflooded by the LAN segment on all links connecting to SPBM other than the one of arrival (e.g., SA = **X** in Fig. 4.8). However, the IS-IS pseudonode (introduced above) used to model the LAN segment can ensure, by using an appropriately high metric for the segment, that the shortest path between B-MACs in the SPBM domain cannot ever transit the LAN segment. Such frames fail RPFC and are therefore discarded.

The outstanding issue is that nodes with unary FDBs could cause frame duplication. By "unary FDB," we mean a node that has a common forwarding table for the entire node, without additional per-port filtering, which is the conventional and necessary structure for a learning bridge. A flood performed by a LAN segment (i.e., VPLS) of a B-MAC unknown to it at one end of the network will create multiple copies of a frame, and then congruent SPF trees across the core and a LAN segment at the other end can defeat RPFC. In Figure 4.8, if DA = **Y** is unknown within the bridged network, its flooding becomes a potential source of multiple frames. Elimination of this phenomenon requires either enhanced filtering or per-port FDBs in devices surrounding LAN segments, as is discussed in more detail now.

In Figure 4.9, the SPF tree to both A and B have multiple congruent components across the core of the network, and the LAN segments at either edge effectively defeat RPFC by making the port of arrival ambiguous. At the top left gateway, A is a valid source (solid path) and B is a valid destination (pecked path); however, that A to B via the top left gateway is not the shortest path, and so the fact that those two addresses are not valid in the **same** frame, is invisible if the FDB is common across the gateway (is unary).

Per-port FDBs suppress this because, with SPBM, in a given B-VID there is only one shortest path from A to B. So if that the bottom left gateway is the only FDB with a valid forwarding entry for MAC address B, the remaining nodes around the switched segment will simply discard the frame.[1]

The simplest way to implement this seems to be a "port validity" bitmap for a destination address entry. Existing filtering defined in 802.1ap [IEEE 2008] states that for a given destination, either all ports

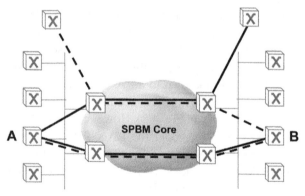

Figure 4.9   LAN segments and frame duplication.

[1] The IETF TRILL/RBRIDGE effort does require a shim for packets transiting share segments such that the ingress to the shared segment can dictate the behavior for load-spreading purposes. As we operate exclusively on edge-based spreading and single shortest path per B-VID, the egress is able to act as filter and remove the obligation on the ingress to be the selector.

or one single port of reception is valid. What is actually required for SPBM unicast is the ability to specify that a selected subset of ports is valid for receipt to a given destination, as a unicast tree is typically a multi-point-to-point structure rooted at the destination.

An alternative technique would be to exploit the "asymmetric VID merge" functionality originally introduced in SPB to allow SPBV to merge multiple SPVIDs into a single VLAN on egress from the SPBV domain. Two VIDs per VLAN would be configured, and with them two FDBs. Ports facing the LAN would use one FDB, and the native SPBM ports would use the other. This allows two different forwarding tables to be set up, one to be used for traffic received from the LAN, the other for traffic received from the SPBM domain. In both cases, the "VID merge" capability allows all traffic to leave the bridge with the VID associated with its egress port, irrespective of its ingress port.

Before leaving the topic of SPBM overlay of bridged Ethernet segments, it is worth noticing that the structure shown previously, with a single Ethernet segment having multiple points of connection with each SPBM region, is not the ideal model in terms of either capacity or resiliency. In particular, if there is a failure, either in the Ethernet segment itself, or one within SPBM that causes routes through the Ethernet segment to change, then reconvergence of the Ethernet segment becomes the gating process.

The alternative is to exploit the ability of link state bridging to use **all** available capacity, which includes multiple parallel paths. When these parallel paths are LAN segments, then the IS-IS pseudonode can be used to model them. This suggests the different deployment model shown in Figure 4.10.

Here, multiple Ethernet LAN segments in the core are used to offer multiple singly connected paths between each SPBM region. Only two such segments are shown, but the principle can be extended to an arbitrary number. This has clear benefits in allowing the capacity of the core segments to scale incrementally, with the allocation of different ECMT algorithms to each B-VID being used to achieve end-to-end load balancing in the usual SPBM manner. The other benefit is more subtle; it is that fault recovery is performed entirely by SPBM itself; either a fault within SPBM or a fault within the Ethernet core is recovered by rerouting within SPBM, which may include the use of another path through the Ethernet core. The need to reconverge the Ethernet core has been eliminated.

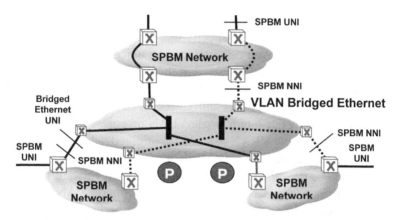

Figure 4.10   Use of multiple Ethernet segments by SPBM overlay.

When overlaying a VLAN bridged core, the core will behave as a broadcast medium for all broadcast and multicast frames. For SPBV, where the VLANs in the core will correspond to the Base VID, the normal operational mode will be trees pruned per core service (i.e., per Base VID). For SPBM, something extra is needed as pruned MAC trees are required within each VID for efficiency. One such solution is 802.1ak [IEEE 2007] MMRP interworking such that SPBM can issue multicast registrations into the bridged core.

Using SPBV over Shared Segments

The control plane aspects of shared segments within SPBV are supported by the same techniques to those described above for SPBM. Within an SPBV region, shortest path trees are constructed using the IS-IS pseudonode to represent the LAN segment(s). Trees crossing a LAN segment all transit the pseudonode and are therefore pruned within an SPBV domain such that the set of LAN gateways collectively has one and only one connection to each bridge in the domain (after excluding other LAN gateways, to prevent reflooding of traffic into the LAN segment by looping through the SPBV domain).

Furthermore, a gateway must only send traffic into the LAN segment that has come from the SPVIDs of bridges which lie on the pruned tree from the IS-IS pseudonode and transiting that gateway (the reverse of the path described above):

- This ensures that traffic flooded into the LAN segment is flooded only once, and using a path that is congruent with the return path from the LAN segment.

Finally, gateways that do not transmit from a particular source into the LAN segment (under the rule above) must block that SPVID on their ports facing the LAN segment, to prevent traffic from that source being reflooded though the LAN segment.

The connectivity set up by IS-IS now "just works." Broadcast or unknown traffic from the SPBV domain is flooded within that domain and also through a single gateway (the closest to the source bridge) into the LAN segment, where it is flooded to all other domains but discarded by other gateways which would return to its originating domain. On egress from the LAN segment, it is flooded by every gateway, but over the pruned trees, so that precisely one copy reaches each bridge in the domain. Path congruence is achieved throughout, and so MAC learning operates correctly in every node transited.

It is worth observing that SPBV's use of MAC learning means that the IS-IS knowledge of SPB domains interconnected by a single LAN segment is actually of little value as the regions do not actually require knowledge of each other in order to operate. This is because all inter-regional paths must transit the LAN segment, and IS-IS connectivity across the LAN segment is not required to disseminate MAC information.

Since SPVIDs are potentially a scarce resource in SPBV, it might be thought that there would be merit in considering each SPBV domain as a region and an IS-IS area its own right, with reuse of SPVID space between regions. However, this is only possible if there is a single LAN segment connecting the SPBV regions, which lacks robustness. If there are multiple parallel LAN segments, IS-IS exchange and topology computations are required in order to allow them all to be used, as in the SPBM case.

Whether either of these SPBV capabilities would be worth placing over VPLS is questionable. VPLS proponents have already adopted and endorsed MAC-in-MAC to improve the scaling of H-VPLS, by eliminating the need for C-MAC learning at core VPLS gateways. It would seem to be retrograde to return to forwarding on C-MAC in the VPLS core, which is what the use of SPBV implies. SPBM appears a much better match to the requirement.

CHAPTER 5

Applications of SPB

Throughout this description, the properties and attributes of SPB have been elucidated by reference to the requirements of key applications. The purpose of this chapter is not to repeat this but rather to draw together and summarize the cardinal points of each application.

SPB IMPLEMENTATION OF MEF SERVICES AND METRO INFRASTRUCTURE

SPB is unique at this moment in terms of offering the MEF connectivity constructs directly in the infrastructure data plane. As observed earlier ("Per-Service-Instance Routing and Forwarding," p. 23), per-service multicast is the most bandwidth efficient way of delivering E-LAN and E-TREE over shared infrastructure, and these multipoint services do not benefit from hierarchy and aggregation. Furthermore, shortest path trees for multicast also have the desirable property that adds, moves, and changes to the set of registered participants do not disrupt service to unmodified participants.

Since E-LINE corresponds to IETF's virtual private wire service (VPWS) with a different transport, and E-LAN with the multicast components turned off is simply a many-to-many unicast VPN, it becomes fairly easy to conclude that SPBM is inherently offering all

802.1aq Shortest Path Bridging Design and Evolution: The Architect's Perspective, First Edition. David Allan and Nigel Bragg.
© 2012 the Institute of Electrical and Electronics Engineers. Published 2012 by John Wiley & Sons, Inc.

the connectivity components that could possibly be wanted by a carrier for use as a "service builder" environment. This is actually very important, as the role of infrastructure in a data/Internet world needs to be reconsidered in terms of the building blocks it is expected to provide.

When compared with "packet transport" approaches for Carrier Ethernet, such as PBB-TE or MPLS-TP driven by a GMPLS control plane, when applied to the metro environment, significant differences emerge. Consider the following:

1. The metro is sufficiently sparsely meshed (typically being formed with dual-homed hub-and-spoke connectivity) that "complete route freedom" is usually restricted to selecting one of the two available choices for a path.

2. Most carriers have used packet techniques to reengineer their operational processes such that logical any-to-any connectivity across their infrastructure exists a priori, as a "trunk" of some form, so that service provisioning can be a single touch to add new endpoints to "point-to-cloud" services, and a touch of just both ends to create new point-to-point services.

Consequently, specific engineering of connectivity at service take-up time is treated as a special and undesirable case. In other words, packet transport solutions with a heavyweight control plane are actually only needed for these demanding special cases.

This can be seen as a natural consequence of the growth in the use of packet transport. When separation between customers' traffic was achieved by inelastic time division multiplexing (TDM) techniques, engineering was essential for both connectivity and efficiency. When separation is achieved by virtualization, the statistical multiplexing of aggregates reduces the need for "fine-grained" efficiency and allows engineering using an "observe and react" model. Only when there becomes a significant mismatch between the installed capacity and the offered traffic matrix does selective engineering of major traffic flows become appropriate.

From this perspective, SPBM can be seen as an excellent match to both the service requirements and modern operational practices of metro networks in particular:

- It offers highly scalable virtualization of the complete suite of fundamental connectivity primitives.

- It offers a "single touch per endpoint" service provisioning model.

- Default any-to-any (trunk) connectivity is installed automatically.

- Techniques have been identified which will allow selective engineering of traffic aggregates in the future. The next chapter describes provisioned connection instantiation (PCI) and other techniques.

- Finally, and a significant operational benefit for carriers, conventional media access control (MAC) learning is used at the client layer, and so the control plane is not required to "bridge" the control protocols of client interfaces together in order to exchange forwarding information.

THE DATA CENTER AND GENERAL ENTERPRISE APPLICATIONS

Traditional enterprise data center designs are based on a simple switching hierarchy, of which typically only the first two layers are layer 2 switches; above that, traffic is funneled into a tree of routers. One reason for this is the spanning tree property of blocking all ports under failure or restoration until the network had reconverged; this makes it imperative to limit the extent of any spanning tree instance. Furthermore, load balancing over multiple paths could formerly only be offered by routers, which was a further motivation to constrain the scale of layer 2. Finally, the only virtualization (subnetting) technique was the VLAN, which greatly limited the freedom with which Internet protocol (IP) addresses could be assigned to servers. Collectively, these considerations drove the development of very large Ethernet switches to maximize the number of servers that could be subtended by a single switch, which was then duplicated for resiliency.

For most enterprises, this model at the time of writing remains adequate. However, this is ceasing to be true for all applications for two reasons:

- There are a small number of enterprises offering Web-based services on a global scale (e.g., Google), where the service sophistication is constrained only by Moore's law and the concomitant requirement for connectivity. Such enterprises may be small in number but are substantial in terms of capacity requirements.
- Under the generic label of "cloud computing," specialist enterprises and service providers are establishing service offerings to allow enterprises to outsource their IT needs, in part or in whole, to multiuser hosted IT facilities.

These large data centers "in the cloud" are becoming the modern "central office." The scale and scope of their internal networking requirements make them the industry driver of layer 2 networking. Such data centers challenge Ethernet switching along three main axes:

- support of virtualization,
- resilience, and
- efficient use of communication resources.

In modern large data center practice, there is much emphasis on the flexible assignment to resources to applications, both to respond to dynamic changes of load per application and also to increase the average utilization of active servers, to limit power consumption. This resulted in the concept of the virtual machine (VM), of which a significant number can be mapped to a physical server. The desire is to allow groups of VMs to be individually mapped to a suitable resource anyway in the data center.

This requires virtualization of connectivity to match the virtualization of servers, and SPBM offers a very natural paradigm. The address of an application component itself is the IP address of the VM on which it is running. The group of components, each running on a different VM and collectively supporting the application, is then an IP subnet, which is exactly mapped to a virtual LAN at layer 2, so the application components can communicate with each other as efficiently as possible, at layer 2.

This virtualization is exactly what SPBM supports; an I-SID for the application subnet is instantiated at all the ports to which a VM is connected, and mapped to a local identifier of the virtual port on the server, typically a C-VLAN identifier. This local binding can be made part of the configuration process for instantiating the VM itself. SPBM then automatically builds the state to join the I-SID instances, wherever they may be in the data center; there is no longer any topological constraint on where the different IPs are placed.

In addition to the configuration ease and freedom, a further immediate impact of the use of SPBM in this way is the potential to massively increase the scale of the data center. This is because the core switches see only the B-MACs of the top-of-rack switches, not the individual server MACs or (even more challenging) the VM MACs.

The resiliency requirements of the data center are addressed by two properties of SPBM's link state routing, which offers attributes greatly in advance of those of a spanning tree:

- Link state routing inherently converges much faster, being based on local calculation over a common topology database, and not a distance vector iterative calculation as a spanning tree is.
- Link state routing has the property that traffic which does not transit any changed component(s) in the topology is unaffected by such a change, unlike a spanning tree, which must block all ports until convergence is complete.

So, not only are convergence times much improved, but faults have local rather than global impact.

The final requirement cited above was the efficient use of communication resources. To exploit the freedom to place VMs anywhere requires throughput across these virtualized clusters, spanning the data center, to match that previously only required locally between physical clusters in a single rack or two. The preferred architectures for this are wide arrays of regularly meshed switches (e.g., fat trees), which offer many alternate paths between any two endpoints. An example of this was shown earlier as Figure 3.14 ("fat tree" switching structure). Such structures would have been inconceivable with spanning tree, but SPBM operates with all ports unblocked and can exploit all paths. The work behind the earlier discussion on "Load Spreading: Equal-Cost Trees" (p. 119) was almost entirely motivated by the application of SPBM to the data center environment, and the reader should revisit this if interested.

In many ways, the campus environment is similar to the traditional hierarchically switched data center except for the sheer density of servers and switches. Traffic is funneled up geographically located trees of aggregation switches and into the campus backbone and WAN routers. Local sensitivity to spanning tree properties determined and still determines the depth of switching deemed acceptable. VLANs are used in small to modest numbers to separate IP subnets and broadcast domains, typically at a departmental or functional level. This model of departmental or functional splitting is based on the assumption that users are similarly organized, by floor or building, which is restrictive.

SPBM brings the same attributes to the campus as to the data center. Switching can be carried right across the campus backbone using rich connectivity and benefiting from the resiliency properties of link state. Finally, the ability to define virtual LANs with the I-SID allows much finer granularity IP subnet definition if and when desired, with a significantly reduced configuration burden when accommodating adds, moves, and changes.

RADIO ACCESS NETWORKS (RANs)

The RAN for "4G" wireless extends significantly the requirements on the backhaul technology over previous generations, and it can also benefit from the properties of SPBM. The architecture of the terrestrial parts of the earlier "2G/3G" wireless networks was conventional:

- It was pure hub and spoke; traffic from many BSS (base stations— towers) was backhauled to the base station controller (BSC), where all the local call-handling intelligence was implemented.
- All the fundamental functionality was carried over TDM circuits.

The requirements imposed by the 4G—LTE—design include

- an "all-IP" end-to-end architecture, with significant call-handling functionality devolved to the NodeB itself (the base station—the BSS in 2G/3G parlance) in order to offload the Mobile Management Entity (MME, which is the MSC in 2G/3G parlance), and

- direct communications between clusters of NodeBs (over the X2 interfaces) in addition to communication between each NodeB and the MME (the S1 interface).

The use of SPBM to manage IP subnets, described above in other contexts, can be reapplied to wireless backhaul. SPBM provides a flexible method with which to manage a virtual layer 2 LAN topology, which can be used to create the required IP subnets between base stations and gateways.

In the RAN environment, each S1 (NodeB to MME) interface can be emulated by a point-to-point LAN. Alternatively, a number of S1 interfaces can be mapped to an E-TREE service, with the further possibility of including more than one gateway to support redundant connectivity at layer 2. The X2 interfaces of each cluster of NodeBs are mapped to an E-LAN service to allow direct communication between them over the shortest available path. By doing this, there is configuration flexibility, and the need to IP route is pushed back to the first point that needs to do so, the MME.

MULTICARRIER CONSIDERATIONS

There are several trends in the industry that collectively will serve to both mandate Ethernet external NNI interfaces and mitigate the issues surrounding such handoffs. This is resulting in the emergence of Ethernet "exchanges" and Ethernet itself as a peering technology of choice.

There are three key trends of note:

The first is "active line access," an initiative of the U.K. regulator Ofcom to open up carrier access to "last mile" facilities. This is intended to circumvent the "bad old days" of colocation cages in central offices by providing back end unbundling of packet-based access as an alternative to requiring direct access to copper or fiber, but still minimizing the amount of backhaul the incumbent needed to perform. Ofcom deliberately chose Ethernet and in particular 802.1ad (Q-in-Q) as the handoff interface in order to leverage the economics of Ethernet. At the time of writing, there was a significant interest among other regulatory bodies in adopting such a model.

The second is the MEF work on the E-NNI such that interconnect, multicarrier OAM and service-level agreement (SLA) guarantees could be achieved between multiple carriers offering Carrier Ethernet services.

The third is the emerging work in the IEEE on a resilient E-NNI, which is not intended to require any control plane peering between the actors to link active topologies; the E-NNI structure should effectively isolate the control planes while providing reliable interconnect. This will permit new entrants to deploy SPB while having the ability to peer with existing Carrier Ethernet deployments to extend services "out of territory."

The fundamental issue that these multicarrier applications will have to address is that of network scale. The 802.1ad (Q-in-Q) is usually found to have adequate identifier space (the two Q-tags) for use on a single interface but is severely challenged to scale to handle a substantial network. SPBM accepts Q-in-Q as a native UNI, and its MAC-in-MAC frame format not only provides identifier space for 16M services but also completely separates the service identifier from any transport constraints. This makes SPBM very well suited to the networking needs behind these unbundled interfaces, especially since it natively supports the E-TREE structures needed for access to multiple servers.

CHAPTER 6

Futures

One of the joys of working on something demonstrably new is the opportunity to explore uncharted territory. SPB has been no exception, where the combination of constraints and the liberating aspects of combining commodity technology with interesting computing algorithms has led us down a number of paths as we sought to expand the scalability, efficiency, and flexibility of SPB.

While this book so far has mostly explored what has been standardized and why it is as it is, this section is far more speculative and suggests directions SPB could go in the future. Some of them have been quite comprehensively thought through, and others are simply pointers in the continuing journey.

FURTHER RESEARCH ON LOAD SPREADING ALGORITHMS

It eventually became apparent that the ability to generate two paths with significant node diversity between any two points in the network would not be sufficient to do a good job of spreading load.

Per-hop load spreading using payload inspection to statistically spread load while preserving flow ordering is the current "gold standard" for routed networks; this is often referred to as equal cost

802.1aq Shortest Path Bridging Design and Evolution: The Architect's Perspective, First Edition. David Allan and Nigel Bragg.
© 2012 the Institute of Electrical and Electronics Engineers. Published 2012 by John Wiley & Sons, Inc.

multipath (ECMP). Most of our examination of the behavior of ECMP "as deployed" over several years lead us to conclude that it actually had a number of undesirable traits:

- it required a lot of additional per packet examination per hop which complicated the forwarding process and expended more energy per packet handled;
- most implementations used the client payload as the source of entropy to maximize the randomization, hence it was the antithesis of "OAM friendly." Consequently, network validation required the operations, administration, and maintenance (OAM) system to somehow impersonate client traffic, which, since it is not necessarily known, can only be an imperfect solution;
- it only did a good job of spreading load in highly regular topologies, and a fault in such a regular topology would perturb the traffic distribution across the network;
- the ability to spread load was limited by the quality of the source of entropy found by payload inspection.

However, the decoupling of unicast and multicast congruence, which would be implied by the application of ECMP to SPB, also had some additional implications on link utilization efficiency for unicast. This arises primarily because unicast tiebreaking could be random, instead of directed to generate minimum cost trees (see the sidebar on tree styles earlier, within "Multicast Mechanisms," p. 78). The consequence is that the number of links over which unicast traffic would be distributed (for any set of equal cost trees [ECTs]) was larger, potentially much larger, than that for SPB with congruent multicast, and hence would be expected to have less likelihood of "hot spots."

Ultimately we are seeing two completely different load spreading use cases. The first is a carrier network, where the topology may be arbitrary, the latency differences between paths may be significant as an artifact of geography and distance, and preservation of the Ethernet service model (e.g., frame ordering) and associated OAM properties is required. The second is the large datacenter/cloud computing facility with a highly regular and richly connected switching architecture, with a large number of equal cost paths between any two servers in the network, but with all switching elements local to a facility.

While the need for ECMP is acknowledged for large datacenter architectures, ECMP cannot be applied to the carrier scenario without deconstructing many of the properties that are considered essential. Therefore, other techniques for load spreading are required that matched carrier sensibilities.

The application of additional intelligence to tiebreaking is the primary degree of freedom available in a single topology (set of link metrics for a network). Although supported by ISIS-SPB, multitopology was discarded for the present as requiring both design and significant extra administration for deployment.

Numerous techniques for incrementally modifying the tie-breaking results local to the set of paths upon which tiebreaking was performed demonstrably failed to preserve the necessary tie-breaking properties. What did work to a point was applying common transforms to the link IDs and reranking the paths. Application of a common transform across the set of ECTs normalized the effect, such that the set of ECTs generated by each transform was self-consistent.

The problem with applying a common transform was that the effect was both pseudorandom and correlated, and as such could be considered to be a variation of the "birthday problem." In practice, it underperformed what simple random node ID assignment would achieve in the tie-breaking algorithm for path identification. For example, if a network has 10 links through a given slice across it between two points, a larger number of paths would be required to be generated to ensure some of the load appeared on each of the links, and the amount of load assigned to each link would be uneven. This was due to the probabilistic nature of link ID transformation.

The attraction of node ID transformation and reranking was that it could be performed in a single pass of the database, and this property conditioned the thinking of the group considering the problem. All the investigation of the "art of the possible" in a single pass highlighted that it was impossible to produce a self-consistent and planar ECT set while attempting to impose dependencies on interim results during that single pass. We could not "change the rules" on the fly and expect the computation to work. Introducing dependencies on interim results of path generation in order to improve overall path selection could only be done at certain points in the path generation process.

In retrospect, the resulting inflation of state, inconsistent spreading of load, and the belief that as long as a large number of diverse paths between any two points were being used then an overall good would be achieved, were specious consequences of that train of thought. In essence, the reasoning which had been followed was that "computation was bad," "anything else was good." This ironically flew in the face of what we simultaneously understood to be the primary disruptive aspect of 802.1aq, that computation was good and the resulting protocol simplification could replace all manner of evil.

An alternative approach which is emerging is the realization that if we are to consider dependencies in path selection in order to improve path selection, the "figure of merit" is no longer how widely the load is spread between any two points in the network. It becomes the evenness with which the total load is spread over the entire network, and this is what we really want to worry about. Hence, the goal after rethinking the entire process has become to optimize the distribution of load in the network when considering the combination of network topology and all sources. In essence, we elevate the path selection dependencies to become part of the state of the network, not just crude diversity between any two points. This leads us down a path where many fewer sets of ECTs are required to achieve efficient network utilization.

The normal problem with computing optimal load distribution in a single pass is that it is an exceptionally hard problem, and made even more so if the result is constrained to a single set of ECTs. If we add the additional constraint of symmetric congruence for the ECT set, the problem is near intractable to solve in any sort of real-time timeframe.

What makes the problem tractable for SPB is that Ethernet allows us to instantiate multiple congruent full mesh topologies from a single set of metrics for any reasonably meshed network. What this then allows us to do is to **incrementally** build sets of ECTs where the placement of subsequent sets considers the placement of all of the sets previously generated, and as such these later sets can be placed to seek out what is anticipated, on the basis of the earlier passes, to be lightly loaded resources.

The technique starts with a first pass through the database to provide a baseline anticipated distribution of load, and generates the well-understood two sets of ECTs derived from low-high tiebreaking from the ranked equal cost path sets.

During the course of computing the first sets of ECTs, we count the number of paths between node pairs that transit each link, referred to as the Ethernet Switched Path (ESP) Count. For the subsequent passes of the database, we additionally filter the set of equal cost paths by first considering the paths with the lowest sum of ESP counts, only reverting to the original lexicographic tiebreaking when there is not a uniquely lowest loaded path. The result is that anticipated link loading, reflected as seeking paths with the lowest ESP counts, dominates tiebreaking for the second and subsequent rounds of ECT path generation. Load is explicitly steered toward the most lightly loaded paths, but still within the context of shortest path. The net result is that the coefficient of variance of link loading across the network is substantially diminished. Each subsequent iteration to generate further ECTs continues to diminish the coefficient of variation but to a lesser extent for each iteration.

If the technique required a number of passes through the database to generate a useful result the technique would not be deployable, as the current state of the art would see this as some multiple of N^2logN computational complexity, or the need for significant retained state after a single computation, plus the state to instantiate the sets of ECTs resulting from each pass. Exploring the technique has shown in practice that for any reasonably meshed network the majority of the benefit occurs in the first two passes (implying four ECT sets), where the coefficient of variation typically diminished some 45% or so from that resulting from the initial low-high ranked path selection. Further iterations would continue to diminish the coefficient of variation, but the effect trails off as the ability to diffuse load is progressively subdivided and at a rate corresponding to the intuitive 1/2, 1/3, 1/4 progression implied by the technique.

An interesting consequence of this approach is that shortest path computation gets a new "knob" to control the distribution of load. To date, playing with shortest path first (SPF) metrics has been a technique fraught with peril, especially in a multipath environment, which tends to take metric manipulation off the table. What this new technique does is permit traffic to be steered away from a hot spot with a simple manipulation of the concept of load associated with a link. If a link has become a hot spot, the count of shortest paths that the link lies on can have a bias factor applied to it in order to push load away from the link during the tie-breaking process.

A key advantage of the technique is that it shifts the burden of multipath from the data plane to the control plane, so infrequent bursts of intense computation substitute for payload inspection and per hop load spreading.

SIDEBAR: *Embedding Payload Entropy within the Packet*

One technique that facilitates OAM testing is to carry entropy information within the layer overhead of a frame. This is a technique employed by MPLS called Flow Aware Transport Pseudo Wires (FAT PWs), where the digest is contained in an entropy label. This principle is currently being proposed for the IEEE 802.1Qbp project on applying ECMP to SPB, in which a new tag format is being considered that would carry an entropy digest in the frame.

The general technique is that the payload is examined at the edge of the network for sources of entropy, and a digest of the entopy is encoded in a known location in the packet header. Intermediate nodes use the entropy information in the packet header as an input into ECMP next path selection.

What this does from an OAM perspective is to allow ECMP to be tested via OAM probing without having to impersonate the payload. What it does not do is allow all possible paths to be proactively verified in real time. The number of entropy values required to authoritatively verify network function is still typically much larger than the actual number of paths in the network and would require a significant number of distinct OAM transactions to test, from thousands to millions.

LAYER 3 INTEGRATION WITH SPBM

To conclude, we expose a whole new capability which the application of Intermediate System to Intermediate System (IS-IS) to Ethernet can bring. In SPBM, the service primitive is the emulated LAN segment. The LAN segment at layer 2 **is** the IP subnet at layer 3, and IS-IS has been routing IP for many years. So, if IS-IS retains its IP personality as well as running SPBM, we have a single control plane with a complete view of both layer 2 and layer 3 topologies, and we can "route at the edge, switch through the core." The SPBM I-SID can now be used to construct a virtual network of IP subnets, and the result is IP-VPN capability as well as the native virtual LAN segments.

In this section we start with an overall introduction to the structures required. We then consider unicast IP forwarding and routing, after which we address how SPBM's virtualization capabilities may be folded in to deliver two IP-VPN models, one lightweight, the other very scalable. We then explain how the native properties of IS-IS allow mixed-capability networks to be built, by which we mean combinations of layer 2-capable nodes, layer 3-capable nodes, and layer 2+3 capable nodes. Finally, we introduce how IP multicast can be added to the repertoire of behaviors by exploiting SPBM's native multicast capabilities.

Introduction to IP/SPBM Integration

The function of IP forwarding with link state bridging is a natural extension of SPBM's optimized SPF calculations, used now to provide a simpler converged protocol for IP unicast and multicast routing. This can be achieved by using the same routing protocol for both layer 2 and layer 3, and when a **single** instance of this protocol is run on each node, a common routing database can be built which binds entities in both layers. This also has the desirable property of allowing virtualized broadcast domains to be arbitrarily partitioned.

This can be achieved because each SPBM node can choose to use one of the IP reachability type-length-values (TLVs) already available in IS-IS. These existing IP reachability TLVs may be announced from other IP-capable SPBM nodes in a network, leaving each SPBM node to decide locally whether to populate a layer 2 route to the announcing node, even if other SPBM nodes decide to construct an IP routing table.

This is possible because any SPBM node, even if it lacks an IP-capable data plane, can relay media access control (MAC)-encapsulated IP traffic, because forwarding in an SPBM network is based on the B-MAC addresses also learned from IS-IS advertisements, and so an IP destination can be associated with the B-MAC address of the node advertising its reachability. A loop-free topology is created based on B-MAC addresses that are derived from the IS-IS System ID portion of the nodes' Network Service Access Point (NSAP) address in the common routing domain; this System ID is used as the router identifier within IS-IS, so it must be unique, and it is 6 bytes in length, allowing it to also be used as the nodal MAC address.

Each node is responsible for sharing its own link state packets (LSPs) with its peers, with TLVs containing information such as its neighbors, locally configured IP subnets, as well as SPBM I-SIDs, and so each TLV type is available for consideration when a node constructs its own filtering database (FDB) state. It is therefore possible for an IP-capable SPBM node, which is building FDB state for end-to-end forwarding based on nodal MACs, to make a single routing decision based on the IP destination of an ingress packet, and construct an Ethernet header to deliver that routed packet as close to the end destination as possible within the SPBM network. This may be viewed as derived from traditional hop-by-hop routing, with the extension that the Address Resolution Protocol (ARP) table of the ingress node has been populated by the layer 2 element of the routing system, not by learning or broadcast query. Furthermore, as indicated above, transit nodes on the route between the IP-routing endpoints need only install forwarding state for the B-MAC associated with the IP destination to establish the path, with no need to perform IP lookup in the data plane.

This method of forwarding IP packets in an Ethernet network has been referred to as SPBM "IP shortcuts." This is a combination of using the PBB-TE-like forwarding behavior of populating the Ethernet FDBs using a control plane rather than learning, together with exploiting the preexisting IP reachability TLVs in IS-IS that carry information about the topological location of IP subnets in a given IS-IS routing domain. The forwarding behavior of SPBM is designed to provide a symmetrical Ethernet switched path between nodes, based on their nodal MAC addresses as a domain-wide unique identifier.

One can summarize the industry's IP routing and switching models succinctly as follows:

- in traditional IP routing, a full IP (longest prefix match) lookup is performed by every router on the path, and the link layer (often Ethernet) is used only to get the frame to the next hop router via any intermediate bridges;
- in MPLS switching, the full IP lookup is performed only on ingress to the MPLS domain, and mapped to the first in a chain of link-local labels which collectively define the path to the egress. The link layer is again used only to get the frame to the next MPLS switch;

- in SPBM integrated with IP, the full IP lookup is performed only on ingress to the SPBM domain. The link layer frame is constructed using the B-MAC address of the egress node of the SPBM domain (the "IP next hop" at the far edge of the SPBM domain), and bridged through to that point without alteration. This achieves end-to-end IP routing over Ethernet forwarding without the core requiring ARP flooding, or reverse path learning.

Another benefit of SPBM is that only a small set of TLVs have been added to IS-IS, and nothing has been taken away; the new TLVs will simply be ignored by pure IP routers. This allows an SPBM node to form an adjacency with a legacy IS-IS router, whereupon all IS-IS nodes seeing the full layer 2 and layer 3 link state topology can calculate a holistic shortest path tree (SPT). This allows SPBM to be introduced into existing networks and provide for an easy migration. When contiguous links are all Ethernet-based, the use of SPBM for IP forwarding becomes a natural alternative to the hop-by-hop methods of IP routing and forwarding. SPBM can make the best use of the native switching links for any services carried on Ethernet or IP.

IP Unicast Address Learning

The capability to advertise IP routes throughout an IS-IS network has existed for many years. The required information for performing a unicast "IP shortcut" can be derived using existing IS-IS TLVs for IPv4 and IPv6. Existing IP Reachability TLVs may readily be combined with SPBM TLVs to bind the required B-MAC information to IP reachability.

In an IS-IS LSP, a node can announce IP routes using narrow and wide value TLVs 128 and 135. In SPBM, the nodal B-MAC address is used as the node's IS-IS System ID, so this appears as the source of every LSP from that node. When forwarding to any of these IP prefixes, this is the B-MAC which is used as the destination address in the Ethernet frame built by the ingress SPBM Backbone Edge Bridge (BEB). This restates the key attribute of this integrated SPBM + IP model; every node runs a single IS-IS instance for both B-MAC and IP layers, and binds LSPs advertising topology and reachability at both layers to a single System ID.

IP reachability information is shared in IS-IS using two styles of TLVs for historical reasons. Old style TLVs 128 (internal IP reachabilities) and 130 (External IP Reachabilities), as well as the new style TLV 135 (Extended IP Reachability), must all be supported on SPBM nodes that announce attached IPv4 addresses.

In the same manner as IPv4, IPv6 can be carried natively in the network without transit nodes needing to be aware that they are forwarding IPv6 packets. IS-IS TLV 236 (IPv6 IP Reachabilities) is used for announcing reachability of IPv6 addresses and is based on the Extended IPv4 TLV 135, and may be used in exactly the same way.

IP-VPN Models

In the preceding sections on IP integration, the reader may have noticed that the virtualization capabilities of SPBM were hardly mentioned:

- the discussion covered layer 2 and layer 3 integration, without any real need for MAC-in-MAC; the exposition was purely concerned with IP over Ethernet,
- which is a "flat," single administration, model.

We now round out the story to include SPBM's virtualization capabilities.

When virtualized services (as instantiated by the SPBM I-SID) are fully exploited, the picture becomes complete and consistent:

- if within a node addressed by a SPBM B-MAC, an I-SID now identifies a virtual routing instance (or VRF),
- and if the IP prefixes associated with that VRF are now advertised in the context of that I-SID, and are only visible to other VRFs also bound to that I-SID, then that I-SID defines an IP-VPN. Specifically, the SPBM Service Identifier and Unicast Address sub-TLV ("New IS-IS TLVs for Link State Bridging," p. 91), defines an SPBM service instance, and so such an SPBM sub-TLV appended to an IP Reachability TLV binds the IP prefixes to that I-SID.

The application described above is the use of a virtual subnet within the core to logically associate a set of edge router functions. Implicit in the description above is that IP prefixes of a particular I-SID or VPN are advertised by the same IS-IS instance as is used to control the backbone connectivity:

- this is the so-called "tagged peer" VPN model.

While the obvious application uses IP routing at the edge and switching in the core, it also becomes possible to consider more interesting architectures. Within a layer 2/layer 3 SPBM domain, the location of IP functionality does not need to surround the core, and virtualization of the subnet can be leveraged to create arbitrary aggregation arrangements, to create, for example, a layer 3 "waypoint" in the middle of the layer 2 core. Any arbitrary set of hosts attached to the network can be organized into a subnet. An umbrella identifier is required to associate multiple I-SIDs with a single layer 3 VPN instance, at which point fully virtualized hybrid layer 2 and layer 3 VPNs can be constructed.

At this point, SPBM's ability to emulate the true duality of layer 2 and layer 3 connectivity models is clearly demonstrated:

- at layer 2, SPBM I-SIDs define a virtualized LAN segment,
- at layer 3, the identical connectivity is seen as a virtual IP subnet.

This is a simple extension to SPBM, and continues to use a single control protocol only; however, this clearly has limited scalability, because in the IS-IS control plane every node has to participate in exchange of reachability information for all endpoints of all VPNs.

However, this approach does include, by definition, multicast VPN constructs, since multicast connectivity is inherent in how an I-SID operates. To use IETF terms, it inherently offers "inclusive tree" (all endpoints) capability per IP-VPN; IP multicast is introduced in the next section. The ability to offer selective trees can also be considered but requires additional logic in the BEBs/ Provider Edges (PEs) to map multicast groups to another I-SID defining a "selective tree."

A more scalable solution can be achieved using the techniques of RFC 4364 [IP-VPN]:

- in which a BGP overlay between BEBs (the "PEs" of RFC 4364) exchanges customer IP prefixes (i.e., per I-SID prefixes), requiring the core SPBM IS-IS to be aware of SPBM network B-MAC addresses only, exactly as when operating a pure layer 2 SPBM network;

- this is the "tagged overlay" VPN model.

This is not just "theoretical"; SPBM has inherited from 802.1ah the forwarding semantics to achieve this capability now:

- the 802.1ah I-TAG structure includes the *Use Customer Addresses (UCA)* flag,
- this, when clear, signals that the C-MAC addresses are to be ignored, and that frame parsing should restart at the Ethertype following the C-MAC fields,
- which in this case would be the customer IP (in the context of the I-SID).

Clearly, this is not as compact a format as could be achieved, and a more efficient frame format can and probably will be defined, but this would be an optimization, and is not an immediate necessity.

IP Multicast

SPBM's integrated multicast capabilities will also allow it to be used as an integrated IP unicast/multicast protocol, by adding a new TLV for IP multicast route propagation, and reusing the optimized all-pairs algorithms of SPBM in place of SPF tree calculation. This TLV will allow a BEB to announce the attachment of a multicast source for a particular group and the I-SID to use for the calculation, or it will announce that a BEB wishes to be a receiver for a source/group (S,G) pair which another BEB has announced. In other words, the IS-IS TLV for IP multicast would be used for binding endpoints which announced a particular multicast address, but the trees needed to install the required forwarding state would be entirely built at layer 2 by SPBM. The advertisement into SPBM of the I-SID associated with the multicast group would be the only mechanism used within SPBM. The means by which an I-SID is associated with an IP multicast group address has not yet been defined.

Any receiver that needs to receive a multicast stream can do so by announcing that it wishes to be a receiver for the multicast group which another BEB has announced that it is transmitting. Any nodes on the path between those BEBs must then use the I-SID to calculate the multicast tree and populate the FDBs as normal. For efficiency, a node could use a single I-SID value for the announcement of a set of IP multicast routes, allowing a single tree to be used for numerous IP multicast streams, which is an entirely foreseeable model when the base I-SID defines the attachment points of a number of sites used by an enterprise.

Any BEB that can also run Protocol Independent Multicast (PIM) may use route redistribution from its PIM neighbors into the SPBM domain. Existing methods of announcing attachment and a request to receive a multicast group by an edge device, such as Internet Group Management Protocol (IGMP), may also be used with SPBM as with any IP multicast routing protocol.

In order to provide IP multicast connectivity, SPBM needs to be able to convey, for each participant, the endpoint's IP source address, the multicast group address which it is announcing, and the I-SID which tandem nodes will use for tree calculation. It must also signify whether the BEB announcing the TLV is the source or wishes to be a receiver, which it does using the same Tx/Rx flags that are used to advertise these I-SID attributes at layer 2. However, in IP multicast, both Tx and Rx bits may only be set if a BEB has multiple ports participating in a specific I-SID, as each individual port participating in an IP multicast is either a transmit or a receive member, but not both.

In some dense multicast scenarios, such as broadcast video distribution, it may be desirable to set up multicast trees independent of the individual services defined by I-SIDs. The tools are available to achieve this:

- the original IP multicast specification, RFC 1112 [IP-MC], defines how an IP host group address can be directly mapped to an Ethernet multicast group address by placing the low-order 23-bits of the IP address into the low-order bits of the Ethernet address.

- The IGMP protocol allows host and IP entities outside an SPBM region to register an interest in receiving a particular IP multicast address.

- The IS-IS-SPB SPBV MAC Address sub-TLV ("New IS-IS TLVs for Link State Bridging," p. 91) has exactly the right parameters to flood this information over the SPB domain. It allows a group MAC address and its desired service (receive-only, transmit-only, or both) to be bound to a VLAN at a node.

This SPBV MAC address sub-TLV was specified to allow the layer 2 multicast registration protocol to register interest in a group MAC at the edge; in this application, IGMP would register interest in an IP multicast address, which is mapped to the corresponding group MAC address and advertised. The anticipated deployment model has a provisioned source, which would advertise the mapped group MAC with the Tx bit set and form the root of an (S,G) tree.

Recipients would translate the IGMP transactions into the advertisement of the group MAC address as receive-only. This would cause the installation of the necessary state to support the tree in the SPB core using identical logic to that employed for per-service multicast tree construction defined by I-SIDs.

It should be noted that the multicast IP address field has 28 bits, and the equivalent MAC field only 23 bits, and therefore there is the possibility of aliasing multiple IP addresses to one MAC address. When deployed, the techniques for ensuring uniqueness of the allocated IP addresses will have to be modified to ensure uniqueness in the 23 bit space.

MULTIAREA

IS-IS for IP supports hierarchical multiarea deployment; therefore, there was an automatic expectation by potential uses that SPB would support such capability too. More rationally, a good multiarea solution permits both the amount of state in a given area, and the convergence time, to be controlled. A single area solution would not be perceived as sufficiently scalable for deployment by a large carrier.

Multiarea for SPBM

Ethernet imposed a number of additional requirements and constraints on the native IS-IS multiarea solution, the key ones being:

- the maintenance of symmetrical path congruency, so the traditional "hot potato" routing performed by IP environments that hide information was not acceptable;
- the inability to summarize unicast MAC address information.

Loop-free operation naturally encouraged us to adopt unmodified the hierarchical routing system which IS-IS already had in its level 1/ level 2 model. If areas were arbitrarily peered, or even just permitted private backdoors between themselves, avoidance of both loops and hot potato forwarding would be substantially more complex.

Since we were already committed to IS-IS for other reasons, its extensibility and layer 3 independence, it was not a difficult decision to go for its multiarea model *in toto*. In this, all traffic between different level 1 areas must transit a single level 2. Any number of area border bridges (ABBs) may be deployed between a level 1 area and the level 2, so capacity is not compromised. Then:

- IF congruence between unicast and multicast, go and return, is maintained,
- AND any service (I-SID) only uses a **single** ABB between levels 1 and 2, then there could never be a loop formed in the transit path from level 1 via level 2 to another level 1, and SPBM's regular loop avoidance mechanisms would guarantee that no loops would form within either level 1 or level 2.

There were a number of possibilities to address the inability to summarize unicast MAC information for SPBM:

1. Use of a C-MAC tandem function at area boundary bridges (ABBs) in order terminate the B-MAC layer and so re-map the B-MACs. This would effectively behave as a MAC-NAT (Network Address Translator) function, with all the state maintaince penalties incurred by NAT functions.
2. Simply filtering which MACs were leaked between areas by taking advantage of I-SID information in the routing system.
3. Hierarchical stacking, producing MAC-in-MAC-in-MAC.

However, it was also recognized that ABBs operating in this regime offered a possibility of condensing the number of multicast trees (as an ABB would be effectively a "root" in the next area for the set of multicast trees that transited it).

While multiarea operation can be achieved with no changes to the existing Ethernet data plane, there is one advantageous change for multiarea that could be considered, which stems from the re-rooting of multicast trees at area boundaries noted above. An advantage of the encoding of multicast MACs described earlier ("Recasting the Group Multicast MAC Address for Shortest Path Trees," p. 66) is that the number of (S,G) trees can be condensed at ABBs by a stateless or "blind" overwrite of the organizationally unique identifier (OUI) portion (which contains the nodal nickname, or SPSource ID) of the DA-MAC of a multicast frame as it moves between areas. It is "MAC-NAT" but in its most simple imaginable form; if the M-bit is set then overwrite the OUI with a fixed value, the nodal nickname of the ABB itself.

An ABB receives frames from all sources in the ISIS level 1 area, but can forward them all into level 2 using its own nickname in the DA-MAC. So the actual information in the header can be interpreted as:

- the source MAC is actual source of the frame (needed for correct binding and learning of C-MAC to source B-MAC at remote nodes);
- the OUI in the multicast DA = root of the current multicast distribution tree (the original source in level 1, the ABB at the ingress to IS-IS level 2, and finally, the ABB at the ingress to level 1 from IS-IS level 2).

ABBs are natural choke points as there is only one egress from a given area for some subset of the (S,G) tree rooted in the parent area. This means that simply overwriting the OUI for multicast frames to respecify the root as corresponding to a particular choke point will condense the number of multicast trees in level 2, and also the number of multicast trees in level 1 rooted in level 2. As FDB consumption is dominated by multicast addresses, this offers a significant scaling benefit for relatively little complexity.

The C-MAC tandem function was considered undesirable due to the large number of C-MACs an ABB may be required to host. Hierarchical stacking was felt to be open to the accusation of "header bloat," and so filtering options were explored.

A solution which actually worked was to specify that ABBs between level 1 and level 2 would auto–elect which ABB represented each level 1 BEB into level 2. The key structures are illustrated in Figure 6.1, and the principles are then described.

Conceptually, this multiarea solution works in a series of steps:

- ABBs between level 1 and level 2 auto–elect which ABB represent each level 1 BEB into level 2, on the basis of the shortest path to a virtual node representing "level 2" which is dual (or more)-homed onto the ABBs.

- The elected ABB would then advertise both the I-SIDs and the nodal B-MACs associated with each BEB it was representing into level 2.

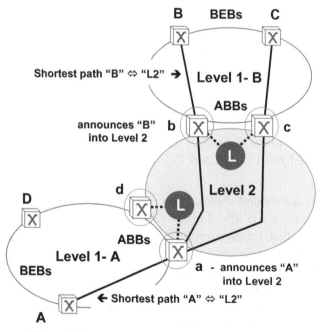

Figure 6.1 A multiarea model potentially used by SPB.

- The nodal B-MACs would be installed throughout level 2 (for OAM).
- ABBs would only advertise I-SIDs from level 2 into a level 1 if the same I-SIDs had already been advertised by that level 1 into level 2.
- Level 2 would only install multicast state and port MACs for an I-SID if that I-SID is advertised into level 2 by ABBs associated with more than one level 1.

One aspect of this needs care in definition. Because the ABB selected by a BEB is determined by the shortest path in level 1, it is entirely possible that a pair or more of BEBs in level 1, each hosting an instance of the **same** I-SID, will choose **different** ABBs to reach level 2. To avoid frame duplication, the following rules are needed:

- all connectivity within a level 1 is provided within that level 1, not in level 2. Therefore, the level 2 multicast trees built off an ABB associated with an I-SID in a specific level 1 must **not** include as sinks any ABBs connected to that same level 1, even if they also advertise that I-SID.
- within a level 1, the multicast tree built off a BEB includes only the ABB which lies on its shortest path to level 2 (this is straight-forward to determine as a by-product of the virtual node used to represent level 2 in the calculation).
- within level 1, the multicast tree built off an ABB includes only the BEBs for which the ABB lies on the shortest path to level 2.

When drawn out, the appearance of the connectivity is point-to-point paths out of level 1 and a spanning tree rooted at the level 1/level 2 boundary. Hot potato forwarding is avoided as the path in level 1 is pinned as point-to-point, and as only one ABB represents each BEB into level 2, there is only one shortest path across level 2 interconnecting any individual pair of BEBs.

One consequence of this was that the shortest interarea path was shortest path in level 1 and shortest path in level 2, but the sum of the shortest paths was not necessarily optimal when considered from a true end-to-end perspective.

There is an alternative (dual) solution, which is to let level 2 drive the ABB designations between pairs of level 1s, by picking the shortest level 2 path between the closest pair of ABBs. I-SIDs are leaked into level 2 as described above, and the I-SID(s) and associated remote B-MACs would be advertised back into level 1 if binding is required, being advertised only from the ABB which was part of the level 2 route between the two level 1s, and thereby telling the level 1 routing system which ABB to use for that service.

It was discovered, however, that there existed pathological corner cases in both approaches, which caused a "clash" of entries in a unary forwarding table, due to a path in one level (2 or 1) to the designated ABB for the other level (1 or 2) transiting another ABB in the same level 1. This problem and its resolution is explained in more detail in the section immediately below.

The Unary FDB Problem at SPBM ABBs

Ongoing investigation of the multiarea solution revealed that there was a problem with ABBs. A common FDB could not always be reconciled with the topology at both level 1 and level 2. Hence, the FDB may be required to support more than one path to a given destination, depending on which level the frame originated from. Figure 6.2 illustrates the manifestation of this problem for the SPB multiarea solution described earlier ("Multiarea," p. 171).

Recall that each BEB chooses the closest ABB to itself in level 1, here ABB 1 for the solid path. The problem arises if the level 2 path transits another ABB (2) on its way to its destination. The problem does not manifest on the outward path, but on the congruent return path (shown as a dashed line in Fig. 6.2). The BEB is advertised via ABB 1 within level 2, but ABB 2 will also have a unicast address for that

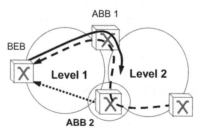

Figure 6.2 Multiarea forwarding and the common (unary) FDB model.

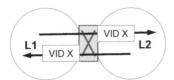

Figure 6.3   SPBM ABB structures.

BEB in level 1 (shown dotted in Fig. 6.2), and the two forwarding entries are in conflict.

The simplest way to address this was to provide distinct B-VIDs to each level, and to provide unidirectional VID translation and merging on egress from the ABB. Hence, a frame arriving in the ABB from (for example) level 1 would arrive tagged with a level 1 B-VID and would use the level 1 FDB. If it left the ABB on a level 2 interface, the B-VID would be translated to the corresponding B-VID in level 2, and so merged with traffic transiting that ABB at level 2. The placement of these functions is summarized in Figure 6.3. This "VID merge" capability is presently being defined and standardized in IEEE 802.1, motivated both by the desire to implement E-TREE in 802.1ad (Q-in-Q) environments, and by the requirements of SPBV itself at region boundaries.

Multiarea Specifics for SPBV

Multiarea for SPBV can be supported by very similar techniques to those described above for SPBM. To achieve this, the gateway bridges perform an asymmetrical rewrite the SPVID in **both directions** across the level 1–level 2 boundary. This process is conceptually identical to the multicast address rewrite described in the previous section for the SPBM case; it re-roots trees at the ABBs, and it summarizes. It further serves to eliminate re-flooding into the original level, as the SPT transiting a gateway, in both level 1 and level 2, does not include any other gateways which attach to the same level 1. Each gateway also has a second level 1 SPVID which it uses for traffic which it locally sources within level 1. This SPVID rewrite automatically solves the problem of the unary FDB in an SPBM ABB, introduced in the previous section in connection with multiarea SPBM. This is because unlike the re-rooting of multicast trees for SPBM, SPVID translation applies to both unicast and multicast.

Within an IS-IS level 1, SPTs are constructed using a "virtual node," as above, which is "connected" to all gateways of that level 1 and represents level 2. This means that gateways appear to source two trees into level 1:

- the one "sourced" by the "virtual node," which is a pruned tree, in which the set of gateways collectively have one and only one connection to each bridge in level 1, and the actual gateway has the shortest path to each included bridge;
- their "own," which is built to all other bridges in the level 1, and used when sending traffic which they locally originate.

The final constraint is that gateways must only ever send traffic into level 2 which came from the SPVIDs of bridges which lie on the pruned tree from the "virtual node," and transit that gateway (as described in the first bullet above); this ensures that traffic flooded into level 2 is sent only once, and using a path which is congruent with the desired return path from level 2.

Within level 2, connectivity using SPTs is built between gateways, except that the tree segments between gateways onto the same level 1 are not joined, to prevent reinjection of traffic back into the originating area via level 2. The same effect can be achieved by building the full mesh of tress, and discarding frames with the SPVID of a peer gateway. The net effect of this is to send traffic which transits level 2 through the shortest path within each level 1 (on the pruned tree), and level 2 then connects the two gateways, using the shortest path within level 2, but the path end-to-end is not necessarily the shortest.

This is now a complete solution. Broadcast or unknown traffic from a level 1 SPBV area is flooded within that area, and also through a single gateway (the closest to the source bridge) into level 2, where it is flooded to all other level 1 areas but discarded by other gateways which would return to its own level 1. On egress from level 2, it is flooded by every gateway, but over the pruned trees, so that precisely one copy reaches each bridge in all other level 1 areas. Path congruence is achieved throughout, and so MAC learning operates correctly in every node transited.

Considerable care would have to be taken in selection of scenarios into which this was deployed. Although the SPVID rewrite and summarization controls VID utilization, all parts of the network are exposed to every MAC address seen. Even if the communities of interest were largely localized within individual level 1 areas, it is likely that aggressive MAC aging policies would be needed to discard MACs flooded but not needed, and there would likely be a further concern of the size of the broadcast domain.

EXTENDED CONNECTIVITY MODELS: SPANNING TREES

The natural baseline connectivity type for SPB is SPTs. In this chapter we range more widely, and show how other connectivity styles may be incorporated into the framework. We first show that traditional spanning trees may also be constructed. We then turn our attention to the coercion of traffic off shortest paths, for traffic engineering purposes, without causing undesirable side effects within the routed system. Finally, we turn the question round, and investigate what other uses SPB's go–return path congruency property may be put to.

Multiple Connectivity Types

There were a couple of motivations for exploring alternative tree construction algorithms for SPB, driven off the common IS-IS database and set of metrics, in particular the ability to use computed spanning trees. The first motivation was a desire to provide more load spreading options in the network without significantly increasing the computational load (this was before we figured out techniques to achieve more than two paths with significant diversity). The other was to gain the scaling benefit of the ability to use (*,G) multicast MAC trees as an adjunct to the (S,G) used by SPTs.

Additional spanning trees were considered to be a small incremental computational load, and they could reuse the synchronization mechanisms for loop avoidance, and the tie-breaking algorithms defined for SPTs. As a spanning tree is a much simpler construct than the mesh of SPTs, the intersection of services on a shortest path would be much simpler to compute and could borrow from the "some pairs shortest path" algorithm optimizations.

The criterion for removing "unsafe" multicast state, to guarantee loop avoidance when neighbors' topology views are not synchronized, is however more complicated than the SPT case just described. This is because the spanning tree is bidirectional; a source on the tree not only sends its traffic toward the root of the tree, but must also send traffic directly to leaves reachable from the root via itself. Determining that the distance to the root has not decreased is an obvious and sufficient test for all nodes which are reached via the root or lie upstream on the path to the root.

Checking distances to the leaves which were previously locally downstream from the root is more complicated. However, we can proceed by extension of the SPT case, where the objective was to prevent unsynchronized inversion of the nodes in the tree order. For the unidirectional SPT, this requires that a node can only keep forwarding without synchronization iff:

- the node does not move closer to the root, and
- the node remains above each of its children.

The spanning tree case is more demanding because of the bidirectional tree, meaning that there are in general **sources** "below" a node (with respect to the spanning tree root), and a node's position with respect to its siblings becomes important.

However, by insisting that a node's distance to the root is **unchanged** if it is to continue forwarding without topology synchronization, the ordering of the complete tree is preserved, and the relative position of siblings as well as parents and children is maintained. Since a loop can only form if node(s) move their positions in the tree hierarchy unnoticed by (unsynchronized with) other nodes, and if such nodes always cease forwarding until synchronized, then loops cannot form. Although a stringent condition, this maintains the important property that trees unaffected by a topology change are not disrupted.

Spanning Trees and the Data Center

Efficient and economic use of networks in data centers continues to present a challenge to the industry. This is principally because of the very large number of path permutations that can exist across a "fat tree" architecture. Applying SPB "as is" to the generation of large numbers of ECT sets is currently computationally intractable, hence there is a debate in the industry around extending ECMP concepts into Ethernet to overcome this problem, but at the expense of symmetric congruency.

A key observation here is that SPB is a general solution, but the problem is highly specialized. Very regular architectures can be handled with much simpler algorithms. For example a fat tree can be served with a radical reduction in computation if IS-IS-SPB computes spanning trees rooted on the top tier nodes, and only reverts to the generalized case in scenarios where a top tier node is known to have failed, or is determined to have residual connectivity to a leaf by transit of another top tier node. In the root failure scenario, spanning tree would normally perform root election (for which only a root located in the top tier would be efficient), whereas if transit of a peer root is possible, the spanning tree computation can be replaced with the normal SPB model, with a tree rooted on every edge device, to diffuse the traffic throughout the surviving resources.

This can be combined with the load spreading algorithms described above such that in a regular architecture the load is uniformly distributed in a fault-free network, and the traffic matrix in the majority of the network is unperturbed in any failure scenarios.

EXTENDED CONNECTIVITY MODELS: NONPLANAR GRAPHS

Provisioned Trees with Routed Backup

Discussions with carriers demonstrated that some wanted very fine control of path placement combined with all the properties of SPBM. Fine-grained path placement completely inverted our model of how SPBM works, which is by aggregating flows when routing, albeit with per service multicast state installation subsequently. Consequently, we chose to view it substantially differently. Such path placement at the level of an individual service required huge amounts of state which would dramatically degrade control plane performance, and hence did not belong in a control plane, but if resilient aggregates could be a viable backup strategy then the role of the control plane would become simply to provide resiliency.

A number of problems quickly emerged. The fundamental issue is that it is impossible to fail over a portion of a multicast tree or a multicast mesh onto precomputed paths, unlike for the case of point-to-point transport, because the required recovery action is completely dependent on the position of the fault in the tree. In other words, recovery of multipoint connectivity with multicast has to be all or nothing.

So the approach was that primary service connectivity would be provided by provisioned trees, configured by a management system using an extension of PBB-TE procedures which would install the required multicast state as well as the unicast B-MACs of PBB-TE. The management system would be assigned as many B-VIDs as was needed to build the required number of different topologies, but any one service would only run in a single one of these topologies. The backup connectivity would be provided by SPBM, which would constantly provide a full and functional FDB based on the current network state. As with any other instantiation of SPBM, multiple ECT algorithms could be run, each assigned to a different B-VID.

A key requirement is the achievement of the "all or nothing" failover of a particular service when a failure occurs. Since the topology is a mesh of SPTs, even a full mesh of connectivity verification sessions between service endpoints does not guarantee to inform all endpoints of a failure, even if such instrumentation was practical from the standpoint of scaling. It would be possible for the management system to cache "link failure impact" records at each BEB, to allow a BEB to map an IS-IS LSP reporting a topology change to a list of local I-SIDs to be switched. However, this involves the significant overhead of installing data related to failure of every link at every BEB, most of which will never be used in the system's lifetime.

However, recall from "Recasting the Group Multicast MAC Address for Shortest Path Trees" that the I-SID appears explicitly as the lower 24 bits of the SPBM multicast address. This means that both nodes at the end of any link can determine **locally** which I-SIDs use the link, simply by inspection of the multicast state installed by the management system when configuring the trees. Indeed, all the housekeeping associated with sending a message to report a failure can be performed in advance, "in case," without significant penalty, because the data is static. So, when a failure occurs, the list of I-SIDs affected by the failure can be immediately flooded by IS-IS to the edge devices in the normal link state fashion. The I-components at such devices would have been provisioned with a "PBB-TE" primary VID and an

SPBM backup VID. When an IS-IS LSP arrived indicating that an I-SID had been affected by failure, the I-component would switch from the primary VID to the backup VID, so restoring connectivity in a manner exactly analogous to a protection switch. An attractive consequence of this use of SPBM is that restoration will occur and reoccur in the presence of multiple faults, as long as connectivity still exists.

One expectation was that there would likely be some similarity between the configured set of available paths and the ECMT set of paths. To mitigate this, individual services could be assigned to a PBB-TE B-VID that in a fault-free state had diverse connectivity from the SPBM backup B-VID. The intention here is that when a failure affected the service on the working B-VID, the backup B-VID itself would be unaffected and stable even while the routing system reconverged.

Topology Modification for Traffic Engineering Purposes

As disclosed so far, SPB has no "knobs" to allow the operator to match the offered traffic to the actual topology and capacity installed (apart from link metrics in IS-IS, which must in general be allocated as a simple function of link bandwidth if distorting and usually undesirable side effects are to be avoided).

In this section, we address this lack of "knobs" and briefly describe a very simple technique, which enables redirection of (significant) traffic between two BEB endpoints off the shortest path, which is the fundamental characteristic of traffic engineering.

The predominant operational model of any-to-any networks such as SPBM is "observe and react," because of the degree to which customer behavior can and does determine the actual traffic flow. Provisioned connection instantiation, or PCI, is a potential first tool for executing the "react" part of this model.

PCI allows a cut-through path to be installed between two nodes, to allow heavy traffic between those points to be diverted onto an optimal path for congestion mitigation. In terms of the forwarding path, for SPBM it is exactly like a PBB-TE trunk installed between the endpoints, and using a different B-VID from any assigned to SPBM. Then the selected I-SIDs can be cut over onto the engineered path by management action reassigning the B-VID on which they run. This is illustrated in Figure 6.4.

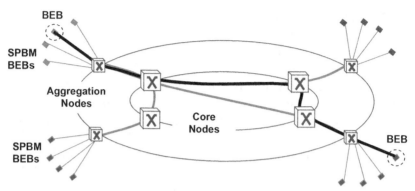

Figure 6.4 PCI: PBB-TE trunk between individual endpoints.

This "just works" for unicast traffic. The significant issue is the handling of the multicast elements of those services. If all traffic between the endpoints is placed on the PCI "trunk," then the multicast tree for the remainder of the service must be rebuilt, and this implies a very tight degree of coupling between IS-IS and PCI, especially under fault conditions.

The alternative is to leave the IS-IS multicast (and unicast) paths intact, and just add the PCI as an overlay. Multicast/unicast congruency is thereby lost, but pseudocongruency (fate sharing) can be maintained by explicit instrumentation of the PCI "trunk" as if it were PBB-TE, using unicast connectivity verification messages, and protection switching of the PCI trunk back to the best-effort tree on failure. In this way, the customer is guaranteed that his connectivity is never worse than measured by C-MAC layer OAM, even if his own OAM never probes the unicast path.

It is possible to create an equivalent capability for SPBV. An Ethernet Virtual Circuit is installed between the selected endpoints, using a VID different from any in service as an SPVID. MAC learning works as normal, with an unknown address being flooded over the tree defined by the ingress bridge's SPVID. When reverse learning from the reply, the edge bridge through which the MAC was reached may be inferred from the SPVID of arrival, and subsequent traffic destined for that MAC may be sent via the PCI VID if one has been provisioned.

Although this technique is possible, it has two attributes which make it undesirable:

- SPBV is envisaged to be the link state controlled successor to traditional bridging, inheriting its highly valued "plug and play" attributes. Traffic engineering of the sort described represents a move into a managed environment which does not fit well with this model.
- These operations do not conform with the Ethernet architecture, in particular with the semantics of VLANs. In such an arrangement, the ingress bridge is looking up a learned MAC, and using that to select the egress VID, as well as the port.

Conclusion

We did not so much set out to do a good job of Ethernet as much as it demanded it of us.

The combinations of restrictions and capabilities imposed by Ethernet, the state of the technology, the service models, and the fundamentals of operation have all tended to constrain the design decisions associated with SPB. This has resulted in a self-consistent approach to unicast and multicast connectivity where in most cases we have been able to specify behaviors with actively desirable properties versus simply "least bad."

The key design constraint which must be and has been maintained is the property of symmetric congruency between unicast and multicast, and the property that there can only be one shortest path to a given destination in a given VID. This means the infrastructure does the best overall job of preserving Ethernet properties when Ethernet links are virtualized over SPB infrastructure, and maximizes the harmonization of component design with existing Ethernet specifications.

This approach results in a number of desirable consequences, which can be seen as evidence for "the virtuous circle."

The first of these is the inheritance of the total body of Ethernet OAM investment. SPB is believed to be the first networking technology for which the full OAM suite was specified and had been implemented before the control plane and the forwarding path were first integrated.

The second of these is the enhancement of Ethernet's scaling properties. The use of the MAC-in-MAC forwarding path provides client–server hierarchy, Ethernet's destination-based B-MAC forwarding model provides O(N) scaling in the fast path, and the adoption of a link

802.1aq Shortest Path Bridging Design and Evolution: The Architect's Perspective, First Edition. David Allan and Nigel Bragg.
© 2012 the Institute of Electrical and Electronics Engineers. Published 2012 by John Wiley & Sons, Inc.

state routing protocol gives control plane scaling properties which are comparable to those of IP.

The third is that once the demands of symmetric congruency had pushed us in the direction of computation as a replacement for multicast signaling, whole new prospects emerged for the application of intelligence to resilience and to load spreading.

In retrospect, we now see a certain inevitability in SPB; once refined and written down, it is somehow "obvious" that it had to be like it is. We hope that some of the discussion has carried over to the reader that it certainly did not seem like that when we started.

References

[802.1Q] "IEEE Standard for Local and Metropolitan Networks, Virtual Bridged Local Area Networks," IEEE Std 802.1Q™ 2005.

[G.8032] ITU-T SG15 Draft Rec. G.8032, "Ethernet Rings Protection Switching," consented February 2008.

[Heinanen] Juha Heinanen, "Intra-area IP Unicast among Routers over Legacy ATM," (draft-ietf-ion-intra-area-unicast-00.txt), July 1997.

[IEEE 2007] Standard for Local and Metropolitan Area Networks Virtual Bridged Local Area Networks—Amendment 07: Multiple Registration Protocol, 2007.

[IEEE 2008] IEEE Standard for Local and Metropolitan Area Networks—Virtual Bridged Local Area Networks Amendment 8: Management Information Base (MIB) Definitions for VLAN Bridges, 2008.

IETF: RFC 4761, "Virtual Private LAN Service (VPLS) Using BGP for Auto-Discovery and Signaling," January 2007.

[IGMP] IETF: RFC 5186, "Internet Group Management Protocol Version 3 (IGMPv3)/Multicast Listener Discovery Version 2 (MLDv2) and Multicast Routing Protocol Interaction," May 2008.

[IP-MC] IETF: RFC 1112, "Host Extensions for IP Multicasting," August 1989.

[IP-VPN] IETF: RFC 4364, "BGP/MPLS IP VPNs," February 2006.

[IS-IS] "Information Technology—Telecommunications and Information Exchange between Systems—Intermediate System to Intermediate System Intra-Domain Routing Information Exchange Protocol for Use in Conjunction with the Protocol for providing the Connectionless-Mode Network Service (ISO 8473)," ISO/IEC 10589, Second edition 2002-11-15.

[IS-IS-L2] IETF: RFC 6165, "Extensions to IS-IS for Layer-2 Systems," April 2011.

[IS-IS-SPB] IETF: draft-ietf-isis-ieee-aq-05.txt, "IS-IS Extensions Supporting IEEE 802.1aq Shortest Path Bridging," March 8, 2011.

[MEF-6] "Metro Ethernet Forum, Technical Specification MEF, Ethernet Services Definitions—Phase 2," April 2008.

802.1aq Shortest Path Bridging Design and Evolution: The Architect's Perspective,
First Edition. David Allan and Nigel Bragg.
© 2012 the Institute of Electrical and Electronics Engineers. Published 2012 by John
Wiley & Sons, Inc.

[Metcalfe] Yogen K. Dalal and Robert M. Metcalfe, "Reverse Path Forwarding of Broadcast Packets," *Communications of the ACM*, 21(12), 1978.

[M-LDP] "Label Distribution Protocol Extensions for Point-to-Multipoint and Multipoint-to-Multipoint Label Switched Paths," IETF RFC 6388, November 2011.

[MMRP] "IEEE Standard for Local and Metropolitan Area Networks: Virtual Bridged Local Area Networks Amendment 7: Multiple Registration Protocol," IEEE Std 802.1ak™ 2007.

[M-OSPF] IETF: RFC 1584, "Multicast Extensions to OSPF," March 1994.

[PBB] "IEEE Draft Standard for Local and Metropolitan Networks, Virtual Bridged Local Area Networks, Amendment 6: Provider Backbone Bridges," IEEE 802.1ah D4.1, February 2008.

[PBB-TE] "IEEE Draft Standard for Local and Metropolitan Networks, Virtual Bridged Local Area Networks—Amendment: Provider Backbone Bridge Traffic Engineering," IEEE 802.1Qay D5.0, January 2009.

[Perlman] Radia Perlman, *Interconnections: Bridges, Routers, Switches, and Internetworking Protocols*, 2nd edition. Addison-Wesley, 1999.

[SPB] "IEEE Draft Standard for Local and Metropolitan Networks, Virtual Bridged Local Area Networks, Amendment 9: Shortest Path Bridging," IEEE 802.1aq D3.0, June 2010.

[SPT-HTzeng] Paolo Narvaez, Kai-Yeung Siu, and Hong-Yi Tzeng, "New Dynamic SPT Algorithm based on a Ball-and-String Model," Infocom 1999.

[Spurgeon] Charles E. Spurgeon, *Ethernet: The Definitive Guide*. O'Reilly, 2000.

[TRILL] IETF: TRILL Working Group, "RBridges: Base Protocol Specification," (draft-ietf-trill-rbridge-protocol-15.txt), January 22, 2010.

[VPLS] IETF: RFC 4762, "Virtual Private LAN Service (VPLS) Using Label Distribution Protocol (LDP) Signaling," January 2007.

During the course of writing, individual 802.1 amendments were rolled up into 802-1Q-REV, and that is the definitive 802.1 standard for all aspects not amended by 802.1aq itself. For ease of access to specific topics, the references to earlier amendments have been retained.

Index

Note: Page numbers in *italics* represent figures.

802.1aq Shortest Path Bridging Design and Evolution: The Architect's Perspective, First Edition. David Allan and Nigel Bragg.
© 2012 the Institute of Electrical and Electronics Engineers. Published 2012 by John Wiley & Sons, Inc.

Printed and bound by CPI Group (UK) Ltd, Croydon, CR0 4YY

27/10/2024

14580240-0003